Information Systems Engineering and Management

Volume 39

The book series "Information Systems Engineering and Management" (ISEM) publishes innovative and original works in the various areas of planning, development, implementation, and management of information systems and technologies by enterprises, citizens, and society for the improvement of the socio-economic environment.

The series is multidisciplinary, focusing on technological, organizational, and social domains of information systems engineering and management. Manuscripts published in this book series focus on relevant problems and research in the planning, analysis, design, implementation, exploration, and management of all types of information systems and technologies. The series contains monographs, lecture notes, edited volumes, pedagogical and technical books as well as proceedings volumes.

Some topics/keywords to be considered in the ISEM book series are, but not limited to: Information Systems Planning; Information Systems Development; Exploration of Information Systems; Management of Information Systems; Blockchain Technology; Cloud Computing; Artificial Intelligence (AI) and Machine Learning; Big Data Analytics; Multimedia Systems; Computer Networks, Mobility and Pervasive Systems; IT Security, Ethics and Privacy; Cybersecurity; Digital Platforms and Services; Requirements Engineering; Software Engineering; Process and Knowledge Engineering; Security and Privacy Engineering, Autonomous Robotics; Human-Computer Interaction; Marketing and Information; Tourism and Information; Finance and Value; Decisions and Risk; Innovation and Projects; Strategy and People.

Indexed by Google Scholar. All books published in the series are submitted for consideration in the Web of Science.

For book or proceedings proposals please contact Alvaro Rocha (amrrocha@gmail.com).

Fadele Ayotunde Alaba · Alvaro Rocha

The Implication of Cyberattacks on Big Data and How to Mitigate the Risk

Springer

Fadele Ayotunde Alaba
Federal University of Education
Zaria, Nigeria

Alvaro Rocha
Department of Information Systems
University of Lisbon
Lisbon, Lezíria do Tejo, Portugal

ISSN 3004-958X ISSN 3004-9598 (electronic)
Information Systems Engineering and Management
ISBN 978-3-031-88569-3 ISBN 978-3-031-88570-9 (eBook)
https://doi.org/10.1007/978-3-031-88570-9

This Springer imprint is published by the registered company Springer Nature Switzerland AG
The registered company address is: Gewerbestrasse 11, 6330 Cham, Switzerland

Competing Interests The authors have no competing interests to declare that are relevant to the content of this manuscript.

Contents

Chapter 1
Introduction

1.1 Background

It has become more important for companies in a variety of industries to be able to gather and analyze massive volumes of data. Big data analytics may provide several advantages to businesses, including enhanced understanding, improved judgment, and more streamlined processes [1]. Big data has become a prime target for hackers who want to get into networks and steal sensitive information because of its widespread use, raising major cybersecurity risks [2]. In several decades, computers and the Internet have become more fundamental in modern society. This has dramatically increased attacks on essential services, including the financial sector, the power grid, government offices, and hospitals [3]. The Internet presents various dangers, including but not limited to targeted attacks, viruses, spam, abuse of system rights, leakage of classified information, exposed vulnerabilities due to lack of maintenance, user indiscretions, and website defacements [4]. The data shows an alarming increase in assaults on these systems and a clear pattern of evolution like these attacks. Although other forms of cybercrime, such as distributed denial of service (DDoS), advanced persistent threat (APT), ransomware, and social engineering attacks, have long been popular, the prevalence of adware, phishing attacks, and Trojans has recently increased. Virvilis reports that there has been an exponential rise in the frequency, severity, and complexity of cyber threats and attacks since 2014 [5].

In the contemporary digitalization era, big data has emerged as a very important asset for both public and private entities. However, this phenomenon has also garnered the attention of nefarious individuals who want to exploit inherent weaknesses within these systems. Cyberattacks targeting large-scale datasets might yield significant ramifications, including breaches of data, infringements upon privacy, financial detriments, misappropriation of intellectual property, data manipulation, and vulnerabilities inside crucial infrastructure [6]. To address these threats, it is imperative to implement heightened cybersecurity protocols, provide comprehensive staff training, enforce strict access control measures, develop robust disaster

F. A. Alaba and A. Rocha, *The Implication of Cyberattacks on Big Data and How to Mitigate the Risk*, Information Systems Engineering and Management 39,
https://doi.org/10.1007/978-3-031-88570-9_1

recovery plans, ensure adherence to regulatory requirements, and foster effective teamwork. Data breaches are a substantial outcome resulting from cyberattacks targeting large-scale datasets [7]. Gaining unauthorized access to confidential data can result in identity theft, financial detriment, and harm to one's reputation. The economic ramifications of cyberattacks are substantial since organizations face financial repercussions from the costs associated with addressing breaches, legal expenses, and penalties imposed by regulatory bodies. Company operational disruptions may lead to financial losses and a decline in market trust [8].

Cyberattacks on large data have significant implications, including infringing intellectual property rights. The misappropriation of trade secrets, proprietary algorithms, and the unauthorized reproduction of products may enable corporate espionage and undermine competitive advantage [9]. Data manipulation is a significant problem due to the potential for hostile actors to disseminate disinformation or false narratives, which may adversely affect public perception, decision-making processes, and electoral outcomes. The potential ramifications of this phenomenon extend to the domains of social trust and stability. The issue of cyberattacks targeting big data raises considerable apprehension over the vulnerability of critical infrastructure. Public safety and security may be compromised when key services, such as electricity grids, transportation networks, and healthcare, are disrupted [10]. To address the potential vulnerabilities associated with assaults on large-scale datasets, organizations need to adopt and enforce advanced cybersecurity protocols. This encompasses sophisticated threat detection systems that use machine learning and artificial intelligence algorithms to promptly detect and address new threats. Regular software upgrades and effective patch management are important in order to address and minimize vulnerabilities. The use of data encryption is crucial for safeguarding data both while it is stored and when it is being sent [11].

The importance of employee training and awareness cannot be overstated when managing cyber hazards. Comprehensive cybersecurity training programmes must educate workers about optimal practices, social engineering dangers, and secure data management. The significance of phishing awareness training cannot be understated, given that phishing continues to be a widespread technique used in cyberattacks [12]. Implementing access control mechanisms and adherence to the least privilege principle are crucial to restricting access to sensitive data and systems. Role-based access control (RBAC) is a security mechanism that guarantees workers access to the information essential for their designated jobs. The notion of least privilege restricts users' access rights and permissions, minimizing the potential attack surface [13]. Implementing disaster recovery and incident response planning is crucial in effectively reducing the adverse consequences of cyberattacks. It is important to have a comprehensive incident response strategy that includes well-defined communication routes and designated teams responsible for responding to incidents. It is essential to establish and implement routine data backup and recovery protocols to effectively assist in restoring data during a security breach or system failure [14].

Ensuring adherence to data protection regulations, such as the General Data Protection Regulation (GDPR) or the Health Insurance Portability and Accountability Act (HIPAA), is paramount in safeguarding the confidentiality of personal data and minimizing any legal liabilities. Implementing cybersecurity insurance may serve as a means to limit potential financial damages in the event of a cyberattack [15]. The significance of collaboration and information sharing cannot be overstated in bolstering cybersecurity endeavors. The act of exchanging threat information with industry peers and cybersecurity organizations serves as a means to remain well-informed about the emergence of potential risks. Public–private partnerships are crucial in promoting and enhancing national and global cybersecurity efforts and resilience [16].

Thus, the ramifications of cyberattacks on large-scale data are substantial and diverse. To address these threats, it is recommended that organizations implement a comprehensive cybersecurity strategy including many components such as heightened security measures, personnel education and training, access control mechanisms, contingency planning for disaster recovery, adherence to regulatory requirements, and fostering collaborative efforts. Implementing proactive and adaptive cybersecurity methods is crucial in safeguarding the invaluable asset of big data and guaranteeing its beneficial impact on society [17].

Cyberattacks on massive data sets might have catastrophic consequences. If hackers get access to a company's big data system, they might compromise its safety and cause financial harm [18]. The repercussions of a cyberattack on a huge data set are significant. Information Theft: A data breach occurs when hackers or other bad actors get unauthorized access to sensitive data. Disclosure of private information, such as names, addresses, and Social Security numbers, as well as trade secrets and intellectual property, may have serious financial, legal, and reputational consequences. When attackers tamper with data within a big data system, it might lead to inaccurate insights and subpar decision-making. The effects might be catastrophic if implemented in sectors where data integrity is critical, such as healthcare, banking, or infrastructure [4]. Service Interruptions: If a cyberattack on a big data system is effective, firms may be unable to access or use their data as planned. Potential implications include substantial monetary loss, business disruption, and dissatisfied consumers [19]. Suppose companies do not take proper precautions to secure their big data platforms and have data breaches. In that case, they might face legal penalties and reputational damage for not meeting data security regulations [20].

Cyberattacks are acts carried out intentionally and maliciously by people or organizations utilising computer systems, networks, or digital devices to compromise data, systems, or services. There are several varieties of cyberattacks, one known as malware assaults, which may take the form of viruses, trojan horses, and ransomware [21]. Emails containing false information and designed to deceive recipients into divulging sensitive data are sent as part of a phishing attempt. Attacks of the kind known as Denial of Service (DoS) and Distributed Denial of Service (DDoS) flood systems with an overwhelming amount of traffic. Attacks known as "Man in the Middle" (MitM) may read and change conversations that are taking place between two parties [22]. SQL injection attacks are designed to take advantage of weaknesses in online applications. Zero-day exploits are designed to take advantage of previously undisclosed vulnerabilities. Social engineering attacks are designed to coerce people into giving information or acting in a certain way [22]. Actions that are deliberate or negligent on the part of workers or partners constitute insider threats. Attacks designed to last for a long time and carried out by knowledgeable foes are known as advanced persistent threats, or APTs. The practise of utilising a victim's computer to mine bitcoins without their permission is known as cryptojacking. There are many different outcomes that may result from cyberattacks, including data breaches, financial losses, reputational harm, service disruptions, and potential threats to national security [23]. In order to protect themselves against these dangers, businesses and people should put in place stringent cybersecurity procedures. This involves performing routine software upgrades, implementing stringent access restrictions, properly educating employees, implementing threat detection systems, and developing incident response strategies. Collective cybersecurity defences may be strengthened by being updated about new threats as they emerge and by exchanging threat information with one another [24].

Companies must take precautions to lessen the likelihood of cyberattacks on big data. Cyberattacks on huge datasets present challenges, and several researchers have written about them and how to mitigate them. This section serves as a summary of some of the most influential books published on the topic. For example, [25]'s "Cybersecurity and Big Data Analytics: An Overview" presents such a summary. This chapter will examine how big data analytics relates to cyber security. It solves the privacy and safety issues that have arisen with the rise of big data. Some methods for reducing risk it clarifies include safe datakeeping, controlled access, and unusual behavior identification. Specifically, "Big Data Security: Challenges, Recommendations, and Solutions" [26]. This book briefly overviews the special security risks in large data contexts and offers some advice for dealing with them. Data security, privacy, intelligence, and permission management are all addressed. From [27]'s upcoming book, "Securing Big Data: A Systematic Literature Review." This literature review looks at studies on safeguarding huge datasets to provide an overview of security procedures and approaches. Two-factor authentication, full-stack encryption, threat monitoring, and other security and privacy measures are discussed.

There is a book coming out in 2020 titled "Big Data Security and Privacy: Challenges and Opportunities" by Chandel et al. This book explores the challenges and opportunities inherent in ensuring the confidentiality and safety of massive databases. Security features such as encrypted data storage, access limits, and anomaly detection are included. An example of such an overview may be found in "Security and Privacy Issues in Big Data: A Comprehensive Overview" [28]. This review study thoroughly examines privacy and security concerns in big data. Data security, access control, secure computing, and the protection of sensitive data are all discussed. It then explains several strategies and techniques for warding off these threats. These books explain the problems with huge data security and provide methods, processes, and recommendations to fix them. Companies that wish to strengthen cybersecurity and protect their big data assets might use them as a starting point.

Because big data plays such an important role in modern business, the threat of cyberattacks cannot be ignored. Understanding the consequences of cyberattacks on big data and implementing effective risk mitigation techniques are necessary to protect vital data assets, keep data integrity, and shield businesses from financial losses, reputational harm, and regulatory compliance violations. By emphasizing cybersecurity, organizations can get the advantages of big data while limiting the associated risks.

1.1.1 Overview of Big Data

The term "Big Data Analysis" may produce conflicting feelings of excitement and dread in the mind of a conventionally trained practitioner in Machine Learning. Existing research endeavors face a substantial challenge presented by the rise of Big Data Analysis, which renders a large portion of the previous work obsolete [29]. This is mostly because algorithms that have been built in the past cannot efficiently manage the enormous amounts of data that are now being processed. In addition, it is possible that these algorithms would be able to appropriately handle the new problems resulting from analyzing big data. In addition, analyzing big data requires a different set of computing talents than those utilized in traditional research, which is something to keep in mind [30]. The investigation of Big Data, on the other hand, is intellectually interesting since it presents a wealth of new difficulties, including both those that have already been discovered and those that have not yet been discovered. Big data analysis, which is bringing a new lease of life to the fields of data mining and machine learning due to the development of new issues, is required due to these new challenges and thus must be used [31].

Nevertheless, what exactly is meant by "Big Data Analysis"? The purpose of this part, as well as the introductory chapter that follows, is to determine the core of Big Data Analysis from the point of view of scientists who work in Machine Learning. We want to examine its potential, which has been extensively stated, for catalysing substantial social reforms, and we want to find out whether it constitutes a considerable divergence from practises that have been used before in the area of machine learning. In addition, we investigate the likelihood that these shifts will be of a more revolutionary or gradual type. Discovering the factors that lie behind the surface of the prevalent excitement is essentially the goal of this investigation. The investigation of a few different ideas about the extraction of Big Data is started, and then it is followed by an investigation into the particular characteristics that are connected to this data. After that, consideration is given to more narrowly focused issues within machine learning. Following a discussion on a large number of notable examples of successful applications in this area, this section comes to a close with a summary of the particular developments that Big Data Analysis has spurred in the fields of study about Machine Learning and Data Mining [10].

It is essential to keep in mind that the field of Knowledge Discovery from Databases, which encompasses data mining, was established in the late 1980s, long before the appearance of Big Data applications and research. The discipline of Machine Learning has been around for quite some time, and throughout that time it has developed a number of algorithms that may be used in data mining operations. Additionally, it has served as a source of inspiration for the creation of more complex and sophisticated solutions. There is a strong overlap between this topic and the field of data mining, particularly from the standpoint of the methodology involved. Others in the academic community have pointed out the differences between traditional data mining and machine learning, while others in the academic community have equated the two ideas [32].

It is still unknown whether or not Big Data has a globally recognized, condensed, and clear definition and whether or not this description is appropriate to machine learning. On the other hand, researchers in the field of machine learning have created an exhaustive list of possible difficulties that may appear in combination with the introduction of Big Data. The academic discourse that may be found in references [23, 24] served as the basis for the tabular representation that was created. On the other hand, to facilitate comprehension and maintain some semblance of order, we have arranged the speeches following the many subject areas. In addition, we enlarged these categories based on our understanding of the field as a whole. It is essential to recognize that the preceding subpart covered a number of the distinctive characteristics of Big Data, which is why you should read it. In light of this, the focus of this part will be narrowed to those aspects of the subject of data mining that are relevant to the application of machine learning approaches [33].

It is essential to keep in mind that the table that has been supplied is just an estimate. In a manner analogous to the difference that may be made between machine learning and data mining, the line that separates the traditional area of data mining and the developing field of Big Data analysis cannot be drawn with absolute clarity. During the early phases of the field, which are often still referred to as traditional data mining, a number of the issues that are now being discussed in the Big Data Analysis category surfaced. It is also important to note that several concerns that were mentioned under the category of Data Mining might perhaps be more appropriately classified under the category of Big Data Analysis, although first, independent research on these matters had started before the emergence of the discipline of Big Data Analysis. At the level of the data gathering, it is also apparent that the distinction between a difficulty involving data mining and a problem involving big data analysis is not recognizable [34], as illustrated in Fig. 1.1.

1.2 Problem Statement

Many methods and best practices have been proposed to lessen the chances of cyber-attacks on big data [3]. Security can be enhanced in many ways, such as through the implementation of stronger authentication and authorization protocols, the use

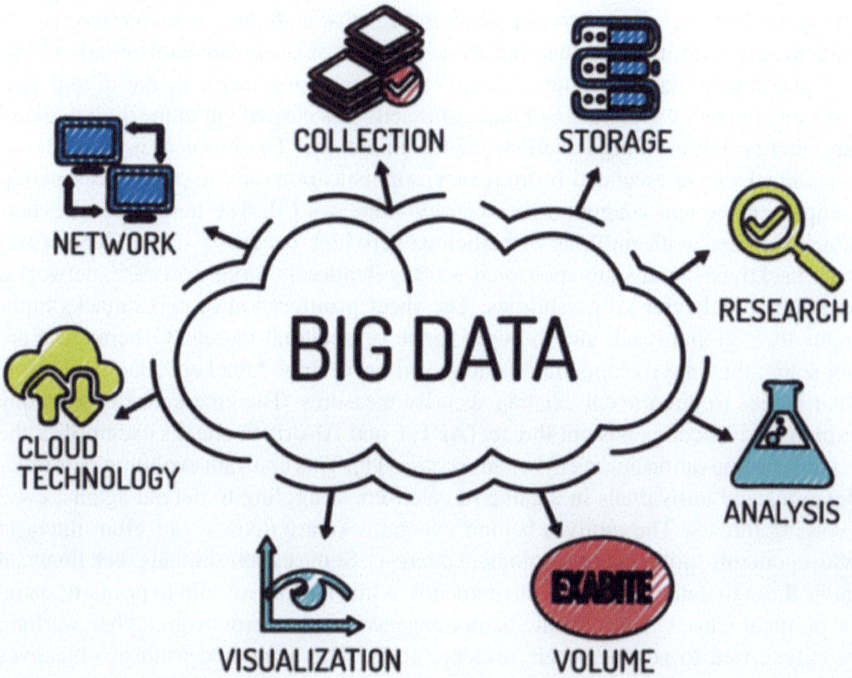

Fig. 1.1 Overview of big data

of stronger data encryption, the implementation of network security measures such as network segregation, the performance of regular security audits and vulnerability assessments, the raising of employee awareness and training, and the implementation of real-time monitoring and incident response capabilities [35]. Abdullahi et al. [3] cite several studies and research books that back up these techniques and provide insight into the challenges and possible solutions for protecting massive amounts of data from assaults. By implementing these safeguards, businesses may lessen the probability that their big data systems will be attacked. "Big data is increasingly important for companies to get insight and make decisions in today's data-driven age [36]. The increasing volume, velocity, and variety of big data makes it more vulnerable to attacks, threatening its confidentiality, integrity, and availability. Protecting big data sets against invasions is challenging because of the need to safeguard sensitive information, prevent data corruption, and maintain stakeholder confidence. Finding a solution to this problem is important for protecting company assets, maintaining smooth operations, and preserving people's right to privacy [18].

The need to protect large amounts of data from cyberattacks is highlighted in this issue statement. It stresses the necessity to use effective methods to prevent data loss, manipulation, and abuse. Data integrity and trustworthiness are essential for organizations and their stakeholders [37]. As big data becomes more important for making decisions and operating organizations, the risk of cyberattacks grows. Cyberattacks on big data may result in various undesirable results, including unauthorized access, data breaches, information manipulation, and the loss of sensitive information [38].

Cyberattacks have become a pervasive and evolving threat in the digital age. They encompass a wide range of malicious activities carried out in the digital realm, targeting individuals, organizations, and even nations. The problem is complex, as cyberattacks have escalated in frequency, sophistication, and impact, necessitating comprehensive and adaptive cybersecurity strategies [3]. The landscape of cyberattacks is vast, with millions of malicious activities occurring daily. Cybercriminals, hacktivists, and state-sponsored actors relentlessly target software, networks, and human behavior vulnerabilities. The sheer proliferation of cyberattacks highlights their global reach and the wide range of potential targets. Cyberattacks are not static; they are dynamic and continuously evolving. Attackers adapt and refine their tactics to circumvent existing security measures. The emergence of zero-day exploits, advanced persistent threats (APTs), and AI-driven attacks exemplifies the relentless innovation among cyber adversaries [1]. This constant evolution puts organizations and individuals in a game of catch-up, struggling to defend against ever-evolving threats. The motives behind cyberattacks are diverse and often intersect with economic, political, or ideological interests. Some cybercriminals seek financial gains through data theft or ransom demands, while hacktivists aim to promote social or political causes. Nation-state actors engage in cyber espionage, cyber warfare, or cybercrime to advance their strategic goals. This wide spectrum of objectives demonstrates that cyberattacks are not confined to a single domain but span various motivations and actors [39].

The consequences of cyberattacks are far-reaching. Data breaches compromise the privacy and security of individuals and organizations, leading to identity theft, financial losses, and reputational damage. The financial ramifications are substantial, with organizations facing costs associated with breach remediation, legal fees, regulatory fines, and public relations efforts [40]. Intellectual property theft harms the victim organization's competitive advantage and facilitates corporate espionage. Manipulation of data and misinformation can undermine trust in media, institutions, and society. Cyberattacks targeting critical infrastructure can disrupt essential services, compromising public safety and national security. To mitigate the risks of cyberattacks, a multifaceted approach is necessary. Enhanced cybersecurity measures include implementing advanced threat detection systems, regularly updating software, and employing robust data encryption [18]. Employee training and awareness programs are crucial to educating employees about cybersecurity best practices and phishing threats. Access control and the principle of least privilege help limit the attack surface. Disaster recovery and incident response planning are essential to guide actions in the event of a cyberattack [41]. Compliance with data protection laws and investing in cybersecurity insurance can mitigate legal and financial risks. Collaboration and information sharing, such as sharing threat intelligence and fostering public–private partnerships, enhance cybersecurity efforts and resilience [39].

These attacks may damage an organization's credibility with its stakeholders and put its finances and reputation at risk. Reducing the risk of assaults on huge data sets is crucial to ensuring the security and resilience of data-driven systems [42]. In conclusion, cyberattacks pose a multifaceted and evolving challenge in the digital age. Their proliferation, evolution, and diverse objectives profoundly affect individuals, organizations, and nations. Mitigating the risks requires comprehensive and adaptive cybersecurity strategies encompassing enhanced security measures, employee training, access control, incident response planning, regulatory compliance, and collaboration [43]. We can only effectively address the complexities of cyberattacks and safeguard against their detrimental consequences through a multifaceted approach.

1.3 Research Rationale

Since many industries are becoming more reliant on big data, it is important to study the possible consequences of cyberattacks on this data and create protections against them. While big data has become a goldmine for organizations, hackers have swarmed it as a key source of important information. Protecting the honesty, privacy, and availability of big data requires understanding the consequences of attacks on this resource. Cyberattacks on large data systems may lead to data leaks, stolen identities, financial fraud, and damaged company reputations. The primary purpose of the research is to get a deeper understanding of the ever-changing cyber threat landscape, to identify and evaluate the many technologies and best practices for securing big data from attacks, and to learn about the particular risks that big data environments face. The research rationale for investigating the implications of cyberattacks on big data is based on several key considerations. Theubiquity of big data in various sectors, such as business, healthcare, finance, and government, makes it a valuable resource for decision-making and innovation. As big data usage continues to grow, understanding the threats it faces and how to protect it is crucial. The pervasiveness of cyberattacks poses a significant challenge, targeting individuals, organizations, and governments worldwide. Understanding the consequences of these attacks, such as data breaches, intellectual property theft, and misinformation campaigns, underscores the importance of devising effective mitigation strategies. Economic and national security are also at risk due to cyberattacks disrupting critical infrastructure, compromising national security, and resulting in substantial economic losses. Research in this area is motivated by the imperative to protect vital systems and resources. Cyberattacks are complex and continually evolving, necessitating ongoing research and innovation in cybersecurity strategies. Legal and ethical considerations surrounding privacy, data protection, and the responsibilities of organizations and governments are also essential.

Global collaboration is crucial to address cyber threats effectively, as they transcend national boundaries. By understanding the nature of cyber threats and the strategies to mitigate them, individuals and organizations can empower themselves to proactively protect their digital assets and privacy.

In conclusion, the research rationale is rooted in the recognition of the critical role of big data in contemporary society and the need to safeguard it from the persistent and evolving threat of cyberattacks. A multidisciplinary approach combining technological innovation, legal frameworks, ethical considerations, and international collaboration is essential to address this complex and multifaceted challenge. Compliance with data privacy and security regulations is mandatory for organizations of any size or field. Researching the aftermath of assaults on massive datasets might provide useful insights about cybersecurity patterns, technologies, and best practices. Cybersecurity may be strengthened, critical data assets protected, business continuity maintained, and customer trust maintained if organizations look at the effects of cyberattacks on big data and create efficient risk mitigation strategies. The study's results and recommendations might improve cybersecurity across various companies and sectors.

1.4 Research Aims

Our present study explores the consequences of cyberattacks on large-scale datasets and recommends possible risk mitigation measures.

1.5 Research Objectives

The research objectives are as follows:

1. To investigate the different forms and consequences of cyberattacks on big data in businesses.
2. To analyze the effect of cyberattacks on large data confidentiality, integrity, and availability.
3. Examining the legal and regulatory structures to safeguard large data from cyber risks.
4. To recommend ideas and recommendations for improving big data security and reducing the danger of cyberattacks.

1.6 Research Questions

The research questions are as follows:

1. What are the different kinds of cyberattacks on big data and their repercussions on businesses?
2. How do cyberattacks affect large data's availability, confidentiality, and integrity?

3. How can we examine the legal and regulatory frameworks for safeguarding large data from cyber threats?
4. What are the recommended solutions and standards for improving big data security and reducing the danger of cyberattacks?

1.7 Research Methodology

The examination of the effects of cyberattacks on big data and the development of methods for risk mitigation often involves a secondary research strategy that consists of the following stages: The research encompasses several essential components, including a comprehensive assessment of relevant literature, the collecting and analysis of data, an investigation of current tools, a comparative analysis, the construction of a framework, the formulation of recommendations, and the validation of findings. The selected research methodology for this study involves a comprehensive examination of existing literature, scholarly publications, academic journals, industrial reports, and other relevant sources. This study aims to thoroughly comprehend the effects of cyberattacks on extensive data systems and the various tactics used to minimize the accompanying hazards. Data collection involves acquiring information from credible sources, such as cybersecurity organizations, government agencies, and experts in the field. Data analysis comprises diverse quantitative methodologies, including statistical analysis and qualitative techniques, such as content analysis. Case studies include the analysis of real-life occurrences in which organizations have faced security breaches on their big data platforms.

The scope of the comparative study is examining and evaluating the methodologies, technologies, and optimal strategies organizations use to mitigate the possible adverse effects of cyberattacks on large datasets. The process of Framework Development involves the construction of a comprehensive framework or model that outlines the fundamental factors, components, and activities involved in mitigating cyberattack risks on extensive datasets.

1.8 Types of Cyberattacks on Big Data

Big data environments are valuable targets for cyberattacks because they store and process massive amounts of data. These attacks can have severe consequences, including data breaches, financial losses, and damage to an organization's reputation [3, 41]. These are some common types of cyberattacks on big data:

1. Data Breaches: Unauthorized access to sensitive data is a common goal of cybercriminals. They may exploit vulnerabilities to access a big data system and steal or manipulate data.

2. SQL Injection: This attack involves injecting malicious SQL queries into web application input fields, which can then be used to manipulate or extract data from a database. In a big data context, this can lead to data compromise.
3. DDoS Attacks: Distributed Denial of Service attacks flood a system with an overwhelming traffic volume, making it inaccessible. In a big data environment, this can disrupt data processing and access.
4. Malware and Ransomware: Malware can be used to compromise big data clusters, and ransomware can encrypt the data, demanding a ransom for its release.
5. Insider Threats: Employees or individuals with access to the big data environment may intentionally or unintentionally compromise data security. They might steal, leak, or manipulate data.
6. Man-in-the-Middle (MitM) Attacks: Hackers intercept communication between systems and manipulate or eavesdrop on data in transit, compromising data integrity and confidentiality.
7. Data Poisoning: In a machine learning and analytics context, attackers may inject false or malicious data into big data sets, affecting the results of data analysis and decision-making.
8. Credential Attacks: Cybercriminals can use phishing or credential stuffing to access user accounts, which may have access to big data resources.
9. Cross-Site Scripting (XSS): This attack involves injecting malicious scripts into web applications, which other users can execute. XSS attacks can compromise data security if a big data platform uses web interfaces.
10. Metadata Attacks: Attackers may target the metadata associated with big data, which can reveal sensitive information about the data itself, its sources, and its usage.
11. Zero-Day Exploits: These attacks target vulnerabilities in software or hardware that are not yet known or patched by the vendor, giving attackers an advantage.
12. Supply Chain Attacks: Attackers may compromise the supply chain of big data tools, infecting them with malware before they are even integrated into the system.
13. IoT Exploitation: Big data often includes data from IoT devices, and attackers may exploit these devices' vulnerabilities to access the larger big data environment.

1.9 Data Breaches and Unauthorized Access

Data breaches and unauthorized access are important security hazards that may have catastrophic repercussions for people, corporations, and even whole countries. Because unlawful access to systems or data often leads to data breaches, these two phrases are frequently used interchangeably. A more in-depth discussion of each of these ideas is presented below:

1. **Breach of Data**

A data breach is an occurrence in which sensitive, private, or protected data is accessed, released, or stolen by an unauthorized person. Examples of this kind of data include credit card numbers, social security numbers, and medical records. Data breaches may be caused by a variety of factors, including cyberattacks, thefts in the real world, and mistakes made by humans. A few important considerations about data breaches.

Data breaches can involve a wide variety of data, including personal information (for example, names, addresses, and social security numbers), financial data (for example, credit card numbers and bank account information), intellectual property and trade secrets, as well as any other information that is regarded as confidential or sensitive.

Hacking, Phishing, Malware, Insider Threats, Weak Access Controls, or Even unintentional Exposure of Data Can Cause Data Breaches Causes: Data breaches can be caused by hacking, phishing, malware, insider threats, or even unintentional exposure of data.

The consequences of a data breach include the potential for monetary losses, harm to the company's image, legal responsibilities, and regulatory penalties. They can potentially result in the affected persons being victims of identity theft, fraud, and other types of cybercrime.

Prevention and Mitigation: To reduce the likelihood of a data breach occurring, organizations need to take preventative steps, such as putting in place robust security procedures, using encryption, establishing access limits and routinely patching vulnerabilities. In the event of a security breach, timely discovery and reaction are absolutely necessary to limit the harm's extent.

2. **Unauthorized Access**

Unauthorized access occurs when a person gains admission to a computer system, network, or data without permission or the required authority. This is the conduct that is referred to as "unauthorized access." Unauthorized access may be an intentional effort to undermine the security of a system, or it might happen accidentally due to a misconfiguration or human mistake. Either way, it is considered to be unauthorized access. Important considerations concerning illegal access:

Unauthorized access may have been motivated by several different activities, including data theft, espionage, sabotage, or simply curiosity.

Methods: To acquire unauthorized access, attackers may use methods such as taking advantage of software flaws, using stolen credentials, launching brute force assaults, or manipulating users via social engineering.

Detection: Detecting unwanted access often calls for the implementation of stringent security measures, such as the use of intrusion detection systems, security logs, and the monitoring of actions that seem to be suspicious.

Prevention: Organizations should adopt robust authentication techniques, access restrictions, and security policies to prevent illegal access. It is also quite important to do regular audits and reviews of access rights.

To protect sensitive data and systems from being compromised, constant monitoring and preventative security measures are required since data breaches and unauthorized access are potential threats. The prevention, detection, reaction, and recovery of an organization's data in the case of a security breach or unauthorized access should be included in any comprehensive security strategy that an organization develops. Additionally, compliance with applicable data protection and privacy rules is essential to data protection and maintaining confidence with important stakeholders.

1.10 Research Significance

Studying the effects of cyberattacks on big data and developing safeguards against them is crucial for several reasons.

1. Safeguarding Private Data: Personal information, financial records, trade secrets, and intellectual property are just a few examples of the types of sensitive and valuable data that may be found in enormous quantities in big data sets. Protecting big data against theft, misuse, and unauthorized access requires understanding the repercussions of cyberattacks on this data.
2. Maintaining Normal Operations During a Cyberattack Financial losses, damaged reputation, and legal repercussions may all result from a cyberattack on a corporation. By recognizing risks and adopting efficient mitigation techniques, businesses may increase their resilience and guarantee business continuity despite cyber attacks.
3. Compliance and legal requirements: Many businesses must follow certain rules and laws regarding data safety and hacking. Cyber risk mitigation in big data may help businesses meet these standards, stay out of hot water, and keep their good name among consumers.
4. Improved Data Governance: Strong data governance standards are essential for effective mitigation methods. Data categorization, access restrictions, encryption, data retention rules, and incident response plans are all aspects of the data governance frameworks that may be established with the aid of this research. The improved data management and security that results from this is invaluable.
5. New ideas and technologies: Being aware of the risks associated with attacks on massive data sets may inspire the development of ground-breaking methods for protecting sensitive information. It encourages the study of innovative approaches to cybersecurity, including machine learning, AI, encryption, and threat intelligence, to cope with the dynamic nature of cyber threats.
6. Collaborating and Sharing Knowledge: Academic institutions, commercial businesses, and government agencies may all benefit from working together to solve the cyber threats big data presents. Information sharing on cybersecurity's achievements, failures, and new advances is a huge boon in the battle against hackers and safeguarding critical data.

7. Public Trust and Confidence: Protecting big data from being compromised by cyberattacks may increase public trust and confidence in businesses and their services. Companies that take preemptive measures in cybersecurity and risk reduction earn the trust of their customers, clients, and other stakeholders.

It is essential for a wide variety of stakeholders and fields for more research on the effects of cyberattacks on big data. It helps protect digital assets, maintains economic stability, protects critical infrastructure and data, ensures public safety, supports innovation and technological advancement, informs the development of legal and ethical frameworks, and spreads knowledge about best practises for cybersecurity. Cybersecurity is an issue that affects people all across the world, and a successful response requires participation from several nations working together. The research results may help countries and organizations work together to find solutions to problems they face in common. The relevance of this study extends to the resilience of digital ecosystems over the long term since an awareness of cyber threats and vulnerabilities is necessary to develop long-lasting remedies. The last point is that research emphasizes technology's ethical and responsible use, encouraging conversations on digital ethics and the social consequences of cyber risks. This study not only helps to solve current problems but also contributes to the continued viability of the digital era in the future. Researchers can help defend digital assets, preserve privacy, maintain economic stability, protect key infrastructure and data, guarantee public safety, create innovation, and promote responsible digital behavior if they grasp the ramifications of cyberattacks on big data. The study contributes to the digital era's long-term viability by doing these kinds of activities.

Companies may improve their data protection efforts and defend the interests of their stakeholders by learning more about the consequences of cyberattacks on big data and how to limit the risk, which can be accomplished by giving good cybersecurity strategy creation top priority.

1.11 Definition of Terms

1. **Big Data**: The term "Big Data" encompasses extensive and intricate datasets that surpass the capacities of conventional data processing technologies. The datasets under consideration are distinguished by their significant volume, rapid velocity, and diverse diversity, sometimes referred to as the three Vs of Big Data. The concept of volume pertains to the storage and analysis of extensive quantities of data, while velocity pertains to the expeditious creation and collecting of data from many sources such as social media, sensors, and gadgets. Variety comprises diverse data kinds, including organised, semi-structured, unstructured, and multimedia data. The concept additionally encompasses two additional factors: integrity, which pertains to the quality and dependability of the data, often encompassing uncertain or incomplete data, and value, which seeks to derive valuable insights, patterns, and knowledge from the data in order to

facilitate improved decision-making and generate business value. Big Data technology and analytics facilitate the processing and extraction of significant insights from vast, rapidly generated, and varied datasets. This empowers organisations to make informed choices based on data [22].

2. **Data Analytics:** Data analytics is a methodical procedure encompassing the scrutiny, refinement, conversion, and elucidation of substantial quantities of data with the aim of revealing significant insights, patterns, and trends that may facilitate informed decision-making and propel commercial achievements. The process encompasses the use of many methodologies, instruments, and advancements to get significant insights from data that is organised, partially organised, or unorganised. The fundamental elements of data analytics include the gathering of data, the refinement and preparation of data via cleaning and preprocessing techniques, the conversion of data into a suitable format, the storage of data, the examination and interpretation of data, the representation of data through visual means, the use of statistical and machine learning models, and the application of business intelligence principles. The process of data collecting entails the acquisition and consolidation of data from several sources, whilst data cleaning and preprocessing procedures are used to assure the veracity and uniformity of the collected data. Data transformation is a process that modifies data to make it more suited for analysis. This might include several techniques, such as normalising, encoding, or scaling the data. Data storage enables the facilitation of effective retrieval and analysis, often using databases or distributed data storage systems. Data visualisation is a method of presenting information in visual representations, such as charts or graphs. On the other hand, statistical and machine learning models are used to develop predictive and prescriptive models, enabling individuals to make educated choices and foresee future trends. Data analytics is extensively used across many domains, including but not limited to business, healthcare, finance, marketing, and scientific research. Its primary objective is to enhance the process of decision-making, find potential prospects, and effectively tackle obstacles via the utilisation of data-driven methodologies. Data Analytics is the process of examining, cleaning, transforming, and modeling data with the goal of discovering useful information, drawing conclusions, and supporting decision-making [44].
3. Hadoop: Hadoop is an open-source framework for distributed storage and processing of Big Data. It allows for the storage and distributed processing of large datasets on clusters of commodity hardware. Hadoop is a free and open-source platform for distributed computing that was developed to store and analyse massive amounts of data while distributing the work over several clusters of commodity hardware. Doug Cutting and Mike Cafarella were the ones who first developed it, and the Apache Software Foundation is the organisation that is responsible for its maintenance. One of the most important parts of Hadoop is the Hadoop Distributed File System (HDFS), which is a kind of distributed file system that offers high fault tolerance, scalability, and data redundancy. MapReduce is a programming methodology and processing engine that enables distributed data processing. It is especially helpful for applications that

need batch processing. Yet Another Resource Negotiator, often known as YARN, is the resource management layer that replaces the older JobTracker and Task-Tracker components. It is responsible for managing and allocating resources to applications that are executing on the Hadoop cluster. The term "Hadoop Common" refers to a collection of utilities, libraries, and application programming interfaces that provide support for other Hadoop modules and include tools for the management and monitoring of Hadoop clusters. Hadoop is an essential component of big data analytics because of its scalability, fault tolerance, and capacity to handle structured and unstructured data. These characteristics have made Hadoop one of the most well-known technologies in the world [45].

4. **NoSQL**: NoSQL, or "not only SQL," is a category of databases designed to handle various types of unstructured, semi-structured, or structured data. NoSQL databases, sometimes referred to as "Not Only SQL" databases, are a distinct category of database management systems that provide an alternative approach to conventional relational databases. These systems are specifically engineered to effectively manage substantial quantities of unstructured or semi-structured data, making them highly suitable for applications involving big data and real-time processing. NoSQL databases possess distinct features and properties that differentiate them from conventional databases. A prominent characteristic of NoSQL databases is their lack of a fixed schema. In contrast to relational databases that need a rigid schema, NoSQL databases provide the capability to store data without a predetermined structure. This characteristic renders them well-suited for managing data that is subject to change and evolution. Scalability is a significant attribute of NoSQL databases. These entities possess significant scalability, exhibiting the ability to scale both vertically and horizontally. This implies that these systems possess the capability to effectively manage substantial quantities of data and may be deployed over several servers or nodes in order to accommodate scalability [46]. The capacity to scale is of utmost importance in big data systems that handle vast quantities of information. NoSQL databases are renowned for their exceptional performance capabilities. These systems have been optimised to cater to certain use cases and data models, hence leading to enhanced query performance. This characteristic makes them very suitable for applications that need rapid data retrieval and processing. One further benefit of NoSQL databases is their ability to accommodate a diverse range of data structures. Various forms of data may be effectively managed by these systems, ranging from basic key-value pairs to intricate network connections. The flexibility of this feature enables developers to choose the most suitable data model based on their own requirements. In addition to their primary functions, NoSQL databases have inherent features that ensure high availability and fault tolerance. The replication of data over numerous nodes guarantees its durability and availability, even in the case of node failures. NoSQL databases are seen a dependable option for applications that need uninterrupted availability. NoSQL databases are often used in a diverse range of applications and settings. These technologies are widely favoured in the realm of online applications, real-time analytics, content

management systems, and the Internet of Things (IoT). Their capacity to effectively manage large levels of data input and output makes them very compatible for such applications. These databases are often used in Big Data environments [47].

5. **Machine Learning:** Machine Learning is a subset of artificial intelligence that focuses on developing algorithms that allow computers to learn and make predictions or decisions based on data, without being explicitly programmed. Machine Learning (ML) is a specialised domain within the science of artificial intelligence (AI) that is dedicated to the advancement of algorithms and statistical models with the aim of enhancing the performance of computer systems in relation to certain tasks. Machine learning algorithms has the ability to acquire knowledge from data and adjust their performance autonomously, rendering them indispensable in many fields including the analysis of extensive datasets, identification of patterns, generation of predictions, and automation of decision-making procedures. The key ideas included in this domain are data, features, algorithms, training, supervised learning, unsupervised learning, reinforcement learning, evaluation, and deployment. Data serves as the fundamental basis for machine learning, whereas features refer to the variables or properties used for the purposes of prediction or categorization [48]. Commonly used algorithms in several fields include linear regression, decision trees, support vector machines, neural networks, and clustering algorithms. The process of training entails providing the machine learning model with labelled data and iteratively modifying its internal parameters in order to minimise the discrepancies between its predictions and the actual results. Supervised learning involves the use of a labelled dataset for training purposes, while unsupervised learning focuses on the analysis of unlabeled data to discern underlying patterns and structures. Reinforcement learning is often used in the domains of games, robotics, and autonomous systems. Evaluation metrics are used to assess the performance of machine learning models. These measures include accuracy, precision, recall, F1 score, and mean squared error. Following the completion of training and assessment processes, machine learning models may be used in practical scenarios to generate predictions or make judgements using novel data. ML finds use in many domains such as Natural Language Processing (NLP), Computer Vision, healthcare, finance, and recommendation systems. With the progression of machine learning, it has become more essential in the automation and optimisation of jobs across many sectors. This advancement has led to improvements in decision-making processes and the facilitation of creative applications [49, 50].
6. **Data Mining**: Data Mining is the process of discovering patterns, trends, and insights in large datasets using techniques from statistics and machine learning. The process of data mining is of utmost importance as it encompasses the extraction of patterns, trends, and important insights from extensive databases via the utilisation of diverse methodologies, algorithms, and statistical approaches. Data analysis is a crucial element in the field, often used to bolster decision-making procedures, detect irregularities, and address intricate issues. The fundamental components of data mining include the gathering of data, the preparation of

data, the exploration of data, and the use of data mining methods. Clustering, classification, regression, association rule mining, and anomaly detection are widely used methodologies with the purpose of uncovering patterns and gaining insights from data. Clustering techniques are used to arrange data points into clusters by considering their similarities, while classification involves assigning data points to pre-established groups or classes based on their distinctive qualities. Regression is a statistical technique used to estimate and predict continuous numerical values by analysing the relationship between input data and the target variable. Linear regression is one commonly used method within the broader field of regression analysis. Association rule mining is a data analysis technique that aims to identify links and linkages among various data components. It is often used in the context of market basket analysis within the retail industry. The primary objective of anomaly detection is to discern atypical or infrequent occurrences within a given dataset, which is crucial for the purposes of detecting fraudulent activities and ensuring the security of computer networks. The evaluation metrics that are particular to each approach are used to measure the efficacy of data mining models. Visualisation tools and approaches are of paramount importance in effectively conveying insights and patterns in a comprehensible manner. Data mining is used in many fields, including business and marketing, healthcare, finance, manufacturing and quality control, as well as online and social media. In summary, the practise of data mining has the potential to uncover significant insights that may be used to influence strategic decision-making, streamline operational procedures, and ultimately improve overall corporate performance. Data analysis is a fundamental component of contemporary data-driven applications, assuming a crucial role in their functioning.

7. **Predictive Analytics:** The term "predictive analytics" refers to a process that makes use of data, statistical algorithms, and machine learning methods in order to make predictions about future outcomes or occurrences on the basis of previous data and patterns. It goes beyond descriptive analytics and focuses on forecasting what may happen in the future rather than just describing what has happened in the past. This procedure gives organisations the ability to make choices that are better informed, so reducing risks and increasing possibilities. Data collection, data preprocessing, feature selection and engineering, feature engineering, predictive models, training and testing, scoring and prediction, model assessment, and deployment are some of the essential components of predictive analytics. The process of collecting data entails accumulating information that is both of high quality and relevance from a variety of sources, including spreadsheets and databases. Data preprocessing includes cleaning and converting raw data, while feature selection and engineering entail picking characteristics that substantially effect predictions. Data preprocessing is a subset of data preparation [51]. For the purpose of creating prediction models, statistical algorithms and machine learning models are applied to historical data. Some examples of these methods and models include regression, decision trees, random forests, neural networks, and support vector machines. In order to assure accuracy and dependability,

training and testing require first teaching models using past data and then evaluating those models using more recent data. The resulting scores are then input into predictive models, which are used to produce forecasts on potential future outcomes or occurrences. Accuracy, precision, recall, F1 score, and area under the ROC curve are some of the measures that are used in the performance evaluation process. When predictive models are deployed in settings that simulate the actual world, it is possible to make predictions or suggestions in real time. It is applicable in a wide range of fields, including as marketing, finance, healthcare, manufacturing, e-commerce, energy, and human resources, amongst others. The purpose of predictive analytics is to improve decision-making by offering insights into future patterns and occurrences. This enables businesses to take preventative action and obtain a competitive edge over other companies [52].

8. **Data Warehouse**: A Data Warehouse is a central repository of data that is collected from various sources and used for reporting and data analysis. A data warehouse is a centralised store of integrated data that is intended to assist operations related to business intelligence and analytics. It brings together data from a variety of sources and reformats it in a standard format, which ensures that the data is consistent and makes it possible to do cross-functional analysis. Because the data in a data warehouse is structured according to certain topics, the information contained within is much simpler to comprehend and examine from a commercial point of view. In addition to this, it maintains previous data, which enables time-based analysis as well as the detection of trends. In most cases, after data has been imported into a data warehouse, it is not updated or modified, which ensures that the data is consistent and stable. Data warehouses are optimised for querying and reporting, and they make use of methods such as indexing and caching to ensure that query performance is lightning quick. They often consist of data marts, which are subdivisions of the data warehouse that concentrate on certain departments or business units and make it simpler for users to obtain data that is relevant to their needs [53]. The data is first extracted from the source systems using ETL methods, then it is transformed, and finally it is loaded into the data warehouse. Data warehouses serve as the basis for business intelligence and analytics tools, which in turn enable companies to explore, analyse, and visualise their data in order to make better business decisions. They provide a single source of truth for reporting and analytics, which guarantees that the data is consistent and reliable, which ultimately leads to improved decision-making. The ability to recognise patterns and to make educated choices based on historical performance is provided to organisations by historical analysis. The accessibility of the data is improved, which makes it possible for a wider variety of users and business units to do their own reporting and analysis. Data warehouses are scalable in the sense that they are able to accommodate a rising number of users and data. In addition to that, they are equipped with advanced safety measures to safeguard critical data. Data warehouses find applications across a wide range of sectors, including retail, healthcare, finance, and manufacturing, among others.

They are necessary for gleaning insights that can be put into action from enormous amounts of data and for driving decision-making inside organisations in a manner that is data-driven [53, 54].

References

1. Seungjin, L., Abdullah, A., Jhanjhi, N.Z.: A review on honeypot-based botnet detection models for smart factory. Int. J. Adv. Comput. Sci. Appl. **11**(6), 418–435 (2020). https://doi.org/10.14569/IJACSA.2020.0110654
2. Mehmood, A., Qadir, A., Ehsan, M., Ali, A., Raza, D., Aziz, H.: Hydrogeological studies and evaluation of surface and groundwater quality of Khyber Pakhtunkhwa, Pakistan. Desalin. Water Treat. **244**, 41–54 (2021). https://doi.org/10.5004/dwt.2021.27913
3. Abdullahi, M., et al.: Detecting cybersecurity attacks in internet of things using artificial intelligence methods: a systematic literature review. Electron **11**(2), 1–27 (2022). https://doi.org/10.3390/electronics11020198
4. Jiang, Y., Zhu, Y., Wu, W., Li, D.: Makespan minimization for MapReduce systems with different servers. Future Gener. Comput. Syst. **67**, 13–21 (2017). https://doi.org/10.1016/j.future.2016.07.012
5. Li, Z.: Accurate digital marketing communication based on intelligent data analysis. Sci. Program. **2022**(2022), 1–10 (2022). https://doi.org/10.1155/2022/8294891
6. Ahmad, M., Amin, M.B., Hussain, S., Kang, B.H., Cheong, T., Lee, S.: Health Fog: a novel framework for health and wellness applications. J. Supercomput. **72**(10), 3677–3695 (2016). https://doi.org/10.1007/s11227-016-1634-x
7. World Health Organisation: WHO Strategic Communications Framework, vol. 2017, no. July. World Health Organisation, p. 56 (2017)
8. Parker, W., et al.: Canadian association of radiologists white paper on de-identification of medical imaging: part 2, practical considerations. Can. Assoc. Radiol. J. **72**(1), 25–34 (2021). https://doi.org/10.1177/0846537120967345
9. Khresna, W.S., Hamsal, M., Furinto, A., Kartono, R.: Implementation of digital transformation to minimize the risk of incidents in the upstream oilfield service quality performance. Sch. J. Econ. Bus. Manag. **8875**(August), 228–234 (2021). https://doi.org/10.36347/sjebm.2021.v08i08.005
10. Zhang, Z., Wen, F., Sun, Z., Guo, X., He, T., Lee, C.: Artificial intelligence-enabled sensing technologies in the 5G/internet of things era: from virtual reality/augmented reality to the digital twin. Adv. Intell. Syst. **4**(7), 2100228 (2022). https://doi.org/10.1002/aisy.202100228
11. Orabi, M., Mouheb, D., Al Aghbari, Z., Kamel, I.: Detection of bots in social media: a systematic review. Inf. Process. Manag. **57**(4). https://doi.org/10.1016/j.ipm.2020.102250
12. Salem, T., Dragomir, M.: Options for and challenges of employing digital twins in construction management. Appl. Sci. (2022)
13. Mannino, A., Dejaco, M.C., Re Cecconi, F.: Building information modelling and internet of things integration for facility management-literature review and future needs. Appl. Sci. **11**(7), 1–18. https://doi.org/10.3390/app11073062
14. Panoff, M., Dutta, R.G., Hu, Y., Yang, K., Jin, Y.: On sensor security in the era of IoT and CPS. SN Comput. Sci. **2**(1), 1–14 (2021). https://doi.org/10.1007/s42979-020-00423-5
15. Liu, L., Wang, P., Lin, J.: ConFlow: Contrast Network Flow Improving Class-Imbalanced Learning in Network Intrusion Detection ConFlow: Contrast Network Flow Improving Class-Imbalanced Learning in Network Intrusion Detection, pp. 0–21 (2022)
16. Coulter, R., Pan, L.: Intelligent agents defending for an IoT world: a review. Comput. Secur. **73**, 439–458 (2018). https://doi.org/10.1016/j.cose.2017.11.014
17. Zupan, L.: 20 key risks to consider by internal audit before 2020. KPMG (2020)

18. Tufail, S., Parvez, I., Batool, S., Sarwat, A.: A survey on cybersecurity challenges, detection, and mitigation techniques for the smart grid. Energies **14**(18), 1–22 (2021). https://doi.org/10.3390/en14185894
19. Hunter, W.C.: J. Smart Tour. **1**(2), 27–36 (2021). http://smarttourism.khu.ac.kr/file/202103/1622686933.pdf
20. Deepa, N., et al.: A survey on blockchain for big data: approaches, opportunities, and future directions. Future Gener. Comput. Syst. **131**, 209–226 (2022). https://doi.org/10.1016/j.future.2022.01.017
21. Heydari, J., et al.: Towards a circular economy for packaging waste by using new technologies: the case of large multinationals in emerging economies. J. Clean. Prod. **11**(1), 438–450 (2019). https://doi.org/10.1016/j.resconrec.2017.08.017
22. Bhattacharya, S., Kumar, P., Maddikunta, R., Pham, Q.: Deep learning and medical image processing for coronavirus (COVID-19) pandemic: a survey. Sustain. Cities Soc. **65**(November), 102589 (2021). https://doi.org/10.1016/j.scs.2020.102589
23. Ceron, J.M., Steding-Jessen, K., Hoepers, C., Granville, L.Z., Margi, C.B.: Improving iot botnet investigation using an adaptive network layer. Sensors **19**(3), 1–16 (2019). https://doi.org/10.3390/s19030727
24. Batool, S., et al.: Lightweight statistical approach towards TCP SYN flood DDoS attack detection and mitigation in SDN environment. Secur. Commun. Netw. **2022** (2022). https://doi.org/10.1155/2022/2593672
25. Al-Garadi, M.A., Mohamed, A., Al-Ali, A.K., Du, X., Ali, I., Guizani, M.: A survey of machine and deep learning methods for internet of things (IoT) security. IEEE Commun. Surv. Tutorials **22**(3), 1646–1685 (2020). https://doi.org/10.1109/COMST.2020.2988293
26. Nesi, P., Pantaleo, G.: Applied sciences IoT-enabled smart cities: a review of concepts , frameworks and key technologies (2022)
27. Porsani, G.B., de Lersundi, K.D.V., Gutiérrez, A.S.O., Bandera, C.F.: Interoperability between building information modelling (Bim) and building energy model (bem). Appl. Sci. **11**(5), 1–20 (2021). https://doi.org/10.3390/app11052167
28. Moradi, J., Shahinzadeh, H., Nafisi, H., Gharehpetian, G.B., Shaneh, M.: Blockchain, a sustainable solution for cybersecurity using cryptocurrency for financial transactions in smart grids. In: 24th Electrical Power. Distribution Conference, EPDC 2019, vol. 2, no. 11, pp. 47–53 (2019). https://doi.org/10.1109/EPDC.2019.8903713
29. Stroumpoulis, A., Kopanaki, E.: Theoretical perspectives on sustainable supply chain management and digital transformation: a literature review and a conceptual framework. Sustainability **14**(8) (2022). https://doi.org/10.3390/su14084862
30. Chiu, M.C., Yan, W.M., Bhat, S.A., Huang, N.F.: Development of smart aquaculture farm management system using IoT and AI-based surrogate models. J. Agric. Food Res. **9**(August), 100357 (2022). https://doi.org/10.1016/j.jafr.2022.100357
31. Ebrahimi, P.: Challenges and Opportunities of Big data and IoT in the Electronic Banking Industry: A Systematic Literature Review, pp. 0–13 (2022)
32. Di Maria, E., De Marchi, V., Galeazzo, A.: Industry 4.0 technologies and circular economy: the mediating role of supply chain integration. Bus. Strateg. Environ. **31**(2), 619–632 (2022). https://doi.org/10.1002/bse.2940
33. Mohammed, M.A., Akawee, M.M., Saleh, Z.H., Hasan, R.A., Ali, A.H., Sutikno, T.: The effectiveness of big data classification control based on principal component analysis. Bull. Electr. Eng. Informatics **12**(1), 427–434 (2023). https://doi.org/10.11591/eei.v12i1.4405
34. Rusch, M., Schöggl, J.P., Baumgartner, R.J.: Application of digital technologies for sustainable product management in a circular economy: a review. Bus. Strateg. Environ. (March), 1–16 (2022). https://doi.org/10.1002/bse.3099
35. Cam-Winget, N., Sadeghi, A.-R., Jin, Y.: Invited—can IoT be secured. In: Proceedings of the 53rd Annual Design Automation Conference DAC'16, pp. 1–6 (2016). https://doi.org/10.1145/2897937.2905004
36. Paka, W.S., Bansal, R., Kaushik, A., Sengupta, S., Chakraborty, T.: Cross-SEAN: a cross-stitch semi-supervised neural attention model for COVID-19 fake news detection. Appl. Soft Comput. **107**, 107393 (2021). https://doi.org/10.1016/j.asoc.2021.107393

37. Sontowski, S., et al.: Cyber attacks on smart farming infrastructure. In: Proceedings of 2020 IEEE 6th International Conference on Collaboration and Internet Computing. CIC 2020, no. December, pp. 135–143 (2020). https://doi.org/10.1109/CIC50333.2020.00025
38. Yoro, R.E., Aghware, F.O., Akazue, M.I., Ibor, A.E., Ojugo, A.A.: Evidence of personality traits on phishing attack menace among selected university undergraduates in Nigerian. Int. J. Electr. Comput. Eng. **13**(2), 1943–1953 (2023). https://doi.org/10.11591/ijece.v13i2.pp1943-1953
39. Zimba, A.: A Bayesian attack-network modeling approach to mitigating malware-based banking cyberattacks. Int. J. Comput. Netw. Inf. Secur. **14**(1), 25–39 (2022). https://doi.org/10.5815/ijcnis.2022.01.03
40. Bures, M., Blazek, P., Nema, J., Schvach, H.: Factors impacting resilience of internet of things systems in critical infrastructure. IEEE Access **21**(5), 2–6 (2022). https://doi.org/10.48550/arXiv.2205.13576
41. Johri, A., Kumar, S.: Exploring customer awareness towards their cyber security in the kingdom of saudi arabia: a study in the era of banking digital transformation. Hum. Behav. Emerg. Technol. **2023** (2023). https://doi.org/10.1155/2023/2103442
42. Movahedi, Z., Hosseini, Z.: A green trust-distortion resistant trust management scheme on mobile ad hoc networks. Int. J. Commun. Syst. (January), e3331 (2017). https://doi.org/10.1002/dac.3331
43. Khan, T., et al.: An efficient trust-based decision-making approach for WSNs: machine learning oriented approach. Comput. Commun. **209**(July), 217–229 (2023). https://doi.org/10.1016/j.comcom.2023.06.014
44. Kehayov, M., Holder, L., Koch, V.: Application of artificial intelligence technology in the manufacturing process and purchasing and supply management. Proc. Comput. Sci. **200**(2019), 1209–1217 (2022). https://doi.org/10.1016/j.procs.2022.01.321
45. Tamboli, S.B., Meit, S.: A novel approach for data intensive caching for big data application using hadoop framework. Access IEEE **2**(March), 1–6 (2015)
46. Bouaziz, S., Nabli, A., Gargouri, F.: Design a data warehouse schema from document-oriented database. Proc. Comput. Sci. **159**, 221–230 (2019). https://doi.org/10.1016/j.procs.2019.09.177
47. Pagán, J.E., Cuadrado, J.S., Molina, J.G.: A repository for scalable model management. Softw. Syst. Model. **14**(1), 219–239 (2019). https://doi.org/10.1007/s10270-013-0326-8
48. Leung, K.M.: Naive bayesian classifier. Polytech. Univ. Dep. Comput. Sci. Risk Eng. (Lecture Notes, 2022). http://cis.poly.edu/~mleung/FRE7851/f07/naiveBayesianClassifier.pdf
49. Devlin, J., Chang, M.W., Lee, K., Toutanova, K.: BERT: pre-training of deep bidirectional transformers for language understanding. In: NAACL HLT 2019—2019 Conference of the North American Chapter of the Association for Computational Linguistics: Human Language Technologies., vol. 1, no. Mlm, pp. 4171–4186 (2019).
50. Agarwal, R., Thapliyal, T., Shukla, S.K.: Vulnerability and transaction behavior based detection of malicious smart contracts. IEEE Commun. Mag. **4**(2), 1–13 (2021)
51. Adeyemi, I.A., Abdul Rahman, M.S., Adeyemi, A.: Maintenance analytics for building decision-making: a literature review. J. Inf. Syst. Technol. Manag. **7**(25), 127–138 (2022). https://doi.org/10.35631/jistm.725010
52. Wang, G., Gunasekaran, A., Ngai, E.W.T., Papadopoulos, T.: Big data analytics in logistics and supply chain management: certain investigations for research and applications. Int. J. Prod. Econ. **176**, 98–110 (2016). https://doi.org/10.1016/j.ijpe.2016.03.014
53. Albara, A.-K., Pradesyah, R.: Power business intelligence in the data science visualization process to forecast CPO prices. Int. J. Sci. Technol. Manag. **2**(6), 2198–2208 (2021). https://doi.org/10.46729/ijstm.v2i6.403
54. Alkhateeb, A., Catal, C., Kar, G., Mishra, A.: Hybrid Blockchain Platforms for the Internet of Things (IoT): a systematic literature review. Sensors **22**(4) (2022). https://doi.org/10.3390/s22041304

Chapter 2
Cyber Attacks

In the contemporary digital landscape, the convergence of big data analytics and the ever-evolving threat landscape has given rise to a significant concern: the implications of cyber attacks on big data. As organizations increasingly harness the power of big data to glean insights and drive strategic decisions, they also find themselves exposed to new vulnerabilities and risks. The potential fallout of cyber attacks targeting big data repositories is far-reaching, affecting not only the security and privacy of sensitive information but also the trust of stakeholders, business continuity, and the overall integrity of decision-making processes [1]. This review delves into the multifaceted implications of cyber attacks on big data systems and explores tools for mitigating these risks. By understanding the intricate interplay between the massive datasets that organizations rely upon and the threats posed by malicious actors, we can better equip ourselves with the knowledge needed to safeguard critical information and ensure the continued advancement of data-driven initiatives [2].

2.1 Big Data Security Platforms

The implementation of Big Data security platforms is of utmost importance for enterprises that are tasked with managing and processing substantial quantities of data. These systems include robust security features to safeguard sensitive data and ensure adherence to regulatory requirements. The fundamental elements and constituents of security platforms for Big Data encompass various components such as access control, encryption, authentication, authorization, data masking and redaction, auditing and monitoring, data loss prevention, firewalls and intrusion detection/prevention systems, security information and event management, tokenization, behavior analytics, compliance and governance, vulnerability assessment and patch management, incident response, data access patterns analysis, machine learning and artificial intelligence, and integration of threat intelligence [3].

F. A. Alaba and A. Rocha, *The Implication of Cyberattacks on Big Data and How to Mitigate the Risk*, Information Systems Engineering and Management 39,
https://doi.org/10.1007/978-3-031-88570-9_2

Access control techniques are used to enforce restrictions on the access of certain data inside the Big Data environment, so ensuring that only people or systems with proper authorization are granted access. The use of encryption is of utmost importance in ensuring the security of data while it is at rest, during its transmission, and during the processing phase. Robust user authentication and permission procedures are essential components of Big Data security. Data masking and redaction procedures are used to safeguard sensitive information by obfuscating or eliminating it from the outcomes or query outputs. Comprehensive auditing and monitoring technologies are used to track and record user activity and system occurrences. Data loss prevention (DLP) solutions are designed to actively monitor and regulate the flow of data in order to mitigate the risk of illegal transfers or breaches. Firewalls and intrusion detection/prevention systems play a crucial role in safeguarding the network and perimeter of the Big Data environment. Security information and event management (SIEM) systems are designed to gather and evaluate security events with the purpose of promptly identifying potential threats in real-time. Tokenization is a data security technique that involves the substitution of sensitive information with tokens, therefore mitigating potential risks [4]. Behavioral analytics solutions have the capability to identify and recognize deviations from normal patterns of behavior, as well as identify possible risks or hazards. The incorporation of compliance and governance components into enterprises facilitates adherence to regulatory frameworks and established norms. Vulnerability assessment and patch management are crucial processes that aim to discover and mitigate security vulnerabilities. Incident response tools and protocols are implemented to effectively handle and mitigate security events. The examination of data access patterns serves to identify instances of unlawful or atypical access. Machine learning and artificial intelligence have the capability to identify and analyze sophisticated security risks. The incorporation of threat intelligence enables firms to remain informed about developing dangers [5].

In general, Big Data security systems provide a comprehensive security framework that enables enterprises to effectively use the advantages of Big Data while ensuring the protection of their data assets against possible security risks.

2.2 Cyberattacks Overview

There is a vast spectrum of cyberattacks, from simple viruses to highly sophisticated malware like cyberweapons. According to [6], the usual suspects—disgruntled employees, hacktivists, and criminal syndicates—are all on the list of most likely cyber-attack origins. This research makes a shocking claim that might indicate a shift in the situation. Conventional wisdom is that the greatest danger to an organization's data security comes from inside, in the form of its employees' actions [7]. The research found that outside assaults are much more dangerous than those launched inside. Several external attackers exist, including state-sponsored actors, hacktivist groups, and lone hackers. These hackers utilize sophisticated malware that is hard to spot even when using cutting-edge security protocols [8]. According to some accounts, the malware used in the Sony attack may have been able to evade current network protections.

Since the release of Stuxnet, a stealthy and devious piece of malware used in an assault on an Iranian nuclear plant, several more variants of this sort of malware have appeared online. Apart from Stuxnet, other instances of stealthy malware include Duqu, Flame, and Red October, all of which fall under the umbrella term "Advanced Persistent Threats" (APTs) [9]. According to [10], an APT has the following characteristics: it is designed to compromise a small number of high-value targets; it uses a specific attack vector, such as a phishing email or a malicious document; it employs a variety of evasion techniques to avoid detection by antivirus and intrusion detection systems (IDS); and it uses command and control techniques to stay one step ahead of security researchers. Because of these characteristics, even highly secure and sophisticated systems often fail to identify APTs. Therefore, finding them requires a lot of manual work and the knowledge of experienced analysts [11]. Thus, an overview of cyberattacks is illustrated in Fig. 2.1.

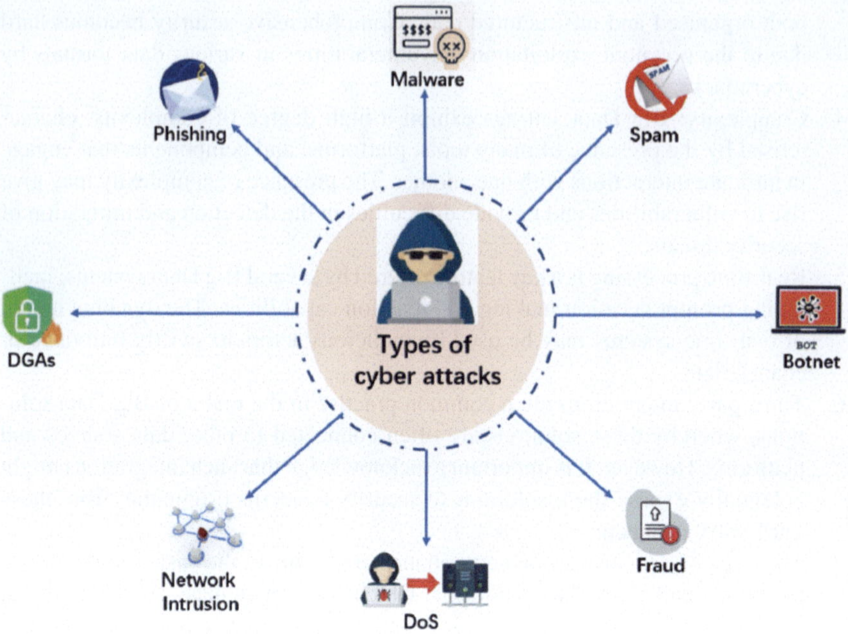

Fig. 2.1 Overview of cyberattacks

2.2.1 *The Increasing Risk of Cyberattacks on Big Data*

The escalating threat of cyberattacks targeting Big Data has become a prominent issue in the contemporary digital environment. As organisations amass and evaluate extensive quantities of data, the potential vulnerability for hackers also increases [12]. The following are few significant elements that contribute to the increasing vulnerability of Big Data to cyberattacks [7, 13]:

1. The proliferation of large-scale datasets in Big Data solutions makes them an enticing target for hackers because to their capacity to analyse and store substantial amounts of data. The substantial quantity of data presents an increased range of possibilities for malicious actors to discover useful information.
2. Big Data encompasses a wide range of data kinds, which often consist of both organised and unstructured data. Cyberattacks have the capability to exploit vulnerabilities present in many data kinds, hence posing a significant challenge in achieving complete security.
3. Big Data encompasses a wide range of data kinds, which often consist of both organised and unstructured data. Comprehensive security becomes hard due to the potential exploitation of vulnerabilities in various data formats by cyberattacks.
4. Complexity: Big Data settings exhibit a high degree of complexity, characterised by the presence of many tools, platforms, and components that engage in intricate interactions with one another. The presence of complexity may give rise to vulnerabilities and provide difficulties in the detection and mitigation of security threats.
5. Real-time processing is a key feature offered by several Big Data systems, facilitating prompt decision-making and reaction capabilities. The dynamic nature of real-time systems may be used by malicious actors to swiftly introduce or change data.
6. Third-party integrations are a common practise in the realm of Big Data solutions, whereby these solutions are often connected to other data sources and platforms. However, it is important to acknowledge that such integrations might potentially expose these solutions to security concerns originating from these third-party connections.
7. The issue of data privacy and compliance arises due to the vast volume of data included inside Big Data platforms, which may encompass confidential and personally identifiable information. Data breaches have the potential to give rise to legal and regulatory challenges, including the General Data Protection Regulation (GDPR), the Health Insurance Portability and Accountability Act (HIPAA), and several other data protection statutes.
8. The use of machine learning and artificial intelligence algorithms in Big Data contexts poses potential threats. These threats manifest in the form of adversarial assaults, which aim to alter the outputs of models or obtain sensitive information.
9. The presence of individuals having authorised access to Big Data systems, sometimes referred to as malicious insiders or workers, may give rise to substantial

security risks. Individuals have the potential to use their granted rights in order to engage in activities such as data theft, fraudulent behaviour, or compromising security measures.

10. Supply chain attacks have the potential to target various components and software used inside Big Data systems. The exploitation of supply chain vulnerabilities in open-source tools and dependencies has the potential to compromise systems.
11. Distributed architectures are often used in Big Data solutions, hence introducing complexities in the realm of security monitoring and threat detection among linked nodes and clusters inside a network.

Organizations must adopt and enforce comprehensive security protocols to address the increasing threat of cyberattacks targeting Big Data [11, 14]. These measures may include a range of strategies or actions.

1. Access Control: It is essential to enforce stringent access controls and authentication systems to limit access to confidential information.
2. Encryption is a security measure involving the transformation of data, both while it is stored and when it is being sent, to safeguard it against unauthorized access.
3. Data anonymization is the process of obfuscating or substituting sensitive data with pseudonyms to mitigate privacy hazards.
4. Monitoring and auditing involves continuously observing and examining data access and user actions to identify any potentially suspicious behaviour.
5. Patch management is the proactive practice of ensuring that software and components are regularly updated in order to effectively mitigate known vulnerabilities.
6. Threat Detection: Implement intrusion detection and prevention systems to detect and address security risks in a real-time manner promptly.
7. Employee Training: This training program aims to provide staff with knowledge and understanding of security best practises, as well as the possible hazards that may arise from using Big Data.

The implementation of a comprehensive data governance policy is crucial in ensuring the maintenance of data quality, integrity, and security. The ever-changing nature of cyber threats necessitates a perpetual and adaptable approach to cybersecurity within Big Data settings. Implementing routine risk assessments and proactive security measures is necessary to protect precious data assets and maintain the confidence of customers and stakeholders [15].

2.2.2 The Role of Big Data in Modern Business and Technology

The advent of the digital era has brought about an unparalleled period characterised by the creation of vast amounts of data, as both people and organisations contribute to its exponential growth. The abundance of data has led to the emergence of the phenomenon generally known as “Big Data.” The concept of Big Data comprises the extensive and heterogeneous datasets that are created across several industries, including but not limited to business, healthcare, and technology [16].

There is a prevailing trend in discussing the potential advantages decision-makers may get from using Big Data. The concept of Big Data has garnered significant attention from several industries, leading to their willingness to allocate financial and operational resources towards realizing the associated business benefits. The use of both Big Data and Little data may lead to significant business advantages. Hence, industries need to engage in thoughtful deliberation over their desired outcomes, followed by a thorough assessment of the availability of pertinent data that might facilitate the attainment of their goals or inform their decision-making processes. In this context, individuals should prioritise enhancing their investment strategies rather than only focusing on increasing their investment amounts. Frequently, the acquisition of pertinent data is necessary in order to attain impartial answers. Prior to the emergence of the Big Data paradigm, statisticians would engage in thoughtful deliberation to determine whether the existing data were sufficient for their objectives or if more data needed to be obtained via an efficient experimental design [17].

For a considerable period of time, statistics has served as a means to address significant research inquiries. The advent of computers has significantly enhanced our capacity to handle increasingly vast datasets, therefore enabling us to address more intricate inquiries to a certain extent. However, it is important to ensure that the data is suitable for its intended purpose, particularly within the context of Big Data. In the age prior to the advent of computers, the process of inverting a 4×4 matrix composed of real numbers was a time-consuming task, however in the present day, this operation has become very easy. Consequently, the field of statistics has seen significant advancements due to the introduction of computers. The advancement of computer statistics research has facilitated statisticians in fitting models of greater complexity compared to prior capabilities via the use of Markov Chain Monte Carlo (MCMC) methods. Furthermore, bootstrap and cross-validation techniques have enhanced the inference process [18].

The perspective posited is that utilizing Big Data presents enhanced prospects, although it is essential to acknowledge the continued relevance of prior knowledge inside the realm of Big Data. It is said that the significance of such knowledge is further amplified. Our perspective asserts that prioritizing the accurate formulation of inquiries has more significance than the availability of suitable data. However, a closely following consideration is the need to have the relevant data to address these inquiries effectively. Big Data is often marketed as a comprehensive solution to all inquiries; nevertheless, we contend this notion is erroneous. In his work, [18] establishes a connection between the advancement of statistics throughout the latter part of the twentieth century and the British-American school's perspective on probability as an objectivistic theory of knowing. According to this perspective, the mathematical notion or model that enables us to comprehend our difficulties must be derived only from observing recurring occurrences without considering any other sources. This observation has significant value within the context of Big Data environments. The first assertion posits that contemporary statistics, as delineated by the authoritative source [19], often known as statistical inference, is a direct derivative of probability theory. The ability to make accurate inferences is contingent upon the development of a comprehensive model that facilitates a thorough understanding of the underlying facts. The second aspect is to the limited knowledge available beyond the mere reiteration of the occurrence, particularly with respect to the use of statistical methodologies. The concept of Big Data suggests the presence of a larger volume of data, however it does not always guarantee a corresponding increase in the amount of meaningful information. The utilisation of Big Data may not be predicated upon the expansion of our existing knowledge or the provision of answers to our significant inquiries. The presence of an abundance of irrelevant information has the potential to mislead or cause confusion on crucial matters, as well as contribute to the generation of erroneous or inaccurate findings. The use of Big Data, which is founded upon a theoretical framework of knowledge discovery, has the potential to enhance our comprehension and expand upon our existing knowledge. The perspective that Big Data provides comprehensive solutions to our pursuit of knowledge, with the only need to identify its presence within the vast dataset, is fraught with peril [20].

The comprehensive resolution of issues pertinent to data custodians is improbable via the use of Big Data unless it has been purposefully designed to accomplish this objective. Most regular datasets primarily gather easily obtainable measures, often consisting of basic administrative data such as revenues and expenditures. These measures are chosen because of their simplicity in measurement and collection or because they are readily available as "open data" that may be downloaded without any associated costs. An illustrative instance pertains to data derived from social networks. Additionally, verifying whether these measurements align with related measurements obtained from the same location is essential, as exemplified in reference [11]. Hence, the advent of Big Data necessitates implementing adequate data management. Enhanced precision may be achieved by implementing a certain degree of aggregation regarding spatial or temporal factors. For instance, one may analyze the mean value obtained during a 5-min interval when data is collected at a frequency of one measurement per minute or by calculating the average of measurements taken inside a defined geographical grid. This approach offers distinct benefits in cases when consecutive measurements of one-minute duration exhibit a significant degree of autocorrelation and nearby measurements effectively capture the same thing. However, this process may lead to a reduction in either spatial or temporal resolution when doing aggregation.

Excessive spatial dimensions or prolonged temporal intervals correspondingly. Hence, it is advisable to include the necessary degree of precision in measurements via the use of suitable data management methods and controls in the measurement procedure.

One of the key challenges associated with sensor networks pertains to where to perform consistency checks on the measurements. This decision involves considering whether to conduct the checks at the location of each sensor before transmitting the information to the root node in the network, thereby disregarding spatial consistency, or to transmit the information to the root node first and subsequently conduct multivariate spatiotemporal consistency checks. The determination of such a selection may not be contingent upon the technique that yields higher levels of precision. However, in the context of wireless solar-operated sensors, this determination may be predicated on considerations about power consumption. However, the precision of measurement will influence the choice of analytical methodology used for data analysis [5, 21].

2.2.3 Examples of Industries Relying on Big Data

Big data has quickly become ingrained in various business sectors, fundamentally altering how these sectors conduct their operations, arrive at decisions, and interact with their consumers. The following is a list of instances of several sectors that depend heavily on big data [22–24]:

1. In medicine, Big Data is used for the diagnosis of patients, the individualization of treatment plans, and the forecasting of disease epidemics. Electronic health records (EHRs) and wearables generate a large amount of data in the healthcare industry.
2. Big Data is helping the financial industry with risk assessment, fraud detection, algorithmic trading, and consumer profiling. The financial sector is one of the most data-intensive industries. Examining market data, transaction records, and sentiments expressed on social media are all quite important.
3. Retail and Online Shopping: Big Data is being used by online retailers to provide personalized suggestions, price optimization, inventory management, and fraud protection. Important things to have are customer data, history of their purchases, and website statistics.
4. Telecommunications: Big Data is used by telecommunications firms to optimize their networks, anticipate the number of customers who will leave, and better focus their marketing efforts. Analysis is performed on customer profiles, call logs, network statistics, and other relevant information.
5. Big Data is essential to operating social media platforms, which use it to facilitate targeted advertising, content suggestions, and trend research. Data about users' activities, interactions, and level of involvement are the essential components.
6. The manufacturing industry makes extensive use of big data for predictive maintenance, quality control, and supply chain optimisation. Information on manufacturing and logistics is gathered together with data from sensors attached to machines.
7. Big data has a number of applications in the energy and utilities industry, including smart grid management, energy usage optimisation, and predictive

maintenance. The information obtained by smart metres and other sensors is of critical importance.

8. Companies in the transportation and logistics industries utilise big data to optimise their routes, estimate their demand, and improve the quality of their vehicle maintenance. Data from GPS devices, information about traffic conditions, and tracking information for shipments all factor into decision-making.
9. Agriculture: Big Data is essential to precision agriculture, which uses the information for crop monitoring, optimizing irrigation, and predicting yields. The data collected from weather stations, sensors, and drones is analysed.
10. The entertainment sector extensively uses Big Data for content suggestion, audience analytics, and targeted marketing. Streaming providers collect information on the viewing habits and activity of their users.
11. Policy-Making, Disaster Relief, and Monitoring Public Health Using Big Data Governments utilise Big Data for policy-making, disaster relief, and monitoring public health. Data from the census, meteorological information, and surveillance all play a part in the decision-making process.
12. In education, Big Data is used by educational institutions for personalised learning, the analysis of student performance, and the distribution of educational resources. The gathering of this information includes student records, data from online learning systems, and assessment results.
13. Big data may assist in creating personalized suggestions, optimizing pricing, and improving the overall consumer experience in the travel and hospitality industries. Data from bookings, customer feedback, and loyalty program insights are all useful sources.
14. Big data is being used in the real estate industry to assist with property appraisal, research of market trends, and investment decision-making. The statistics on housing transactions, location information, and features of the properties themselves are vital.
15. The monitoring of climate change, the tracking of animals, and the evaluation of impacts on the environment are all areas that may be improved by using Big Data, which both organizations and academics use. Data gathered from sensors, satellite photography, and weather stations are all considered.

These are just a few examples, and the effect of big data continues to grow across various industries, leading to more informed decision-making, greater consumer experiences, and increased operational efficiency.

2.3 Type of Tools Available to Solve the Problem of Cyber Attacks on Big Data

This section will discuss the existing tools used in threat detection.

2.3.1 Identity and Access Management (IAM)

Protecting contemporary businesses' digital assets and private data relies heavily on IAM, an essential component of cybersecurity. IAM is a broad framework for controlling the authentication and authorization of users to protect sensitive information [25]. It includes procedures like user provisioning, authentication, authorization, and access control to restrict access to data and services to those who can see them. Single Sign-On (SSO), Private Access Management (PAM), Authentication, and Authorization are the cornerstones of IAM [26]. Data breaches, identity theft, and unauthorized access to sensitive information are some of the many cybersecurity issues IAM solves. It ensures that only authorized people may access the system and that their actions are constantly checked. Increased safety, conformity with rules and regulations, a better user experience, and easier administration are just some of the upsides of a well-implemented IAM system. Organizations may find it difficult to use IAM due to concerns around scalability, user uptake, and system integration. New security flaws necessitate that IAM systems be routinely patched and upgraded [27].

IAM is crucial to contemporary cybersecurity tactics, guaranteeing that only authorized users can access priceless digital assets. IAM solutions allow businesses to improve their security, meet regulatory requirements, and provide customers peace of mind. IAM will remain a cornerstone in securing critical information and warding off cyber dangers as firms embrace digital transformation [28]. Regarding cyber defense, IAM is a critical framework that protects digital identities and administers access to internal resources. It is crucial to keep sensitive information safe, follow the rules and regulations, and keep unwanted people out of sensitive areas [29, 30].

Amazon Web Services (AWS) is a well-known corporation that uses IAM to protect its users better and provide them with a better overall service. By providing clients with safe and granular management of their resources and data in the cloud, IAM at AWS is a crucial component of AWS's security architecture [31]. Customers may use IAM to establish and manage user identities through the AWS Management Console, including users, groups, and roles, each with its permissions based on their specific duties and responsibilities. Password-based authentication, multi-factor authentication (MFA), and connection with other identity providers like Active Directory are just some of the authentication techniques IAM offers [32]. Users, groups, and roles inside IAM are granted access to just those resources for which they have been granted permission. Integrating single sign-on (SSO) with external identity providers streamlines access control and improves the user experience. In addition to user authentication and authorization, IAM offers privileged access management (PAM) to control sensitive data access [33]. AWS's use of IAM bolsters customer security, compliance, scalability, auditability, and adaptability. With IAM's authentication, authorization, and access control features, businesses can monitor user access, enforce security standards, and keep their cloud environment safe. Amazon Web Services (AWS) successful implementation of identity and access management

(IAM) demonstrates how enterprises may utilize this technology to improve their security posture and protect their digital assets in the current digital ecosystem [22].

The Windows Operating System (OS) is very popular in the United Kingdom because of its intuitive design, powerful functions, and ability to work with various hardware and applications. Windows' longevity, adaptability, business-focused features, and compatibility all contribute to the OS's widespread use [19]. By releasing Windows 10 Pro and Enterprise with enhanced features like BitLocker encryption, Group Policy management, and Windows Update for Business, Microsoft has adapted the Windows OS to meet the varying requirements of modern businesses [22, 33]. The widespread use of the Windows operating system has resulted in developing a sizable community of independent software developers and information technology providers that provide their products and services to enterprises. The uniform design concepts, straightforward interface, and simple usage increase productivity and save training time for new users [34]. The Windows operating system and the Microsoft Office suite of programs work together to provide a uniform setting for doing business. The Start Menu and Taskbar, Cortana and Search, File Explorer, Windows Update, and support for touchscreen devices are just a few of the user-friendly elements of the Windows operating system [3]. With these updates, Windows OS is more equipped to meet the needs of UK and international enterprises than ever before. Windows OS continues to develop to suit the changing requirements of organizations in the UK and elsewhere via regular upgrades and enhancements [35].

2.3.2 *Symmetric Data Encryption*

This section will explore how you may use GNU Privacy Guard (GnuPG) to keep your data safe and private. Modern cybersecurity would be impossible without symmetric data encryption, which allows for secret communication and prevents sensitive data from falling into the wrong hands. Due to its speed, efficiency, and simplicity, it is widely used for a wide range of purposes, including the encryption of data in transit, the protection of sensitive data in databases, and the protection of sensitive information on storage devices [36]. However, maintaining the efficiency and safety of symmetric encryption requires efficient key management and key exchange protocols. Even with the constant development of cyber dangers, symmetric encryption has shown to be an indispensable part of any serious data security plan [36, 37]. Symmetric data encryption, also known as secret-key or private-key encryption, encrypts plaintext data using a single secret key and decrypts ciphertext using the same key. The fundamental premise of symmetric encryption is that both parties must possess the same key to communicate securely [30].

GNU Privacy Guard, or GnuPG, is open-source encryption software that uses cryptographic methods to protect users' privacy and data. Developed as a free and open-source alternative to the commercial Pretty Good Privacy (PGP) encryption software, GnuPG is licensed under the GNU General Public License (GPL). Its support for symmetric and asymmetric encryption enables secure data transfer and storage. GnuPG's digital signature features make it possible to generate and validate digital signatures, guaranteeing the genuineness and integrity of transmitted data [38].

This encourages openness, community participation, and the ongoing strengthening of the software's protections and capabilities. GnuPG is available for various platforms, such as Microsoft Windows, Apple macOS, Linux, and Unix-like operating systems. Users may feel safe knowing their data is protected regardless of their platform [19]. Safeguarding private conversations and files from prying eyes, GnuPG guarantees a secure and encrypted exchange of information. Its primary usage is to send and receive encrypted emails and share files between users [39]. GnuPG's digital signature verification features allow people and businesses to create verifiable digital identities and use those identities to digitally sign communications and other documents using private keys [30]. Thus, GNU Privacy Guard (GnuPG) is an extremely robust and flexible encryption software vital to protecting personal information in the digital era. It is widely used by people, corporations, and government agencies to prevent eavesdropping and other forms of data theft because of its strong encryption, portability, and open-source nature. Because it puts users in charge of their data protection and digital identity verification, GnuPG remains a vital instrument in the armory of cybersecurity measures [3, 40].

As cyber risks and insider threats have become more sophisticated, the old security approach that offers broad rights to users upon network access is insufficient in preventing unauthorized access and data breaches. With the Zero Trust Policy in place, all network users, devices, and sections are assumed to be malicious unless proven otherwise [41]. Factors including user ID, device status, geolocation, and activity are used to determine who has access. Access permissions are dynamically altered depending on real-time risk assessments, and user authentication and authorization are performed in real-time [42]. The Zero Trust concept works effectively with the military-grade encryption technology Advanced Encryption Standard (AES) 256. The 256-bit key size of the AES 256 symmetric encryption method makes it very difficult for hackers to crack. Data encrypted with AES 256 is unreadable to anybody without the decryption key [43]. Adding AES 256 encryption to the Zero Trust architecture is a great way to safeguard sensitive data and provide users peace of mind. Only authorized users may access encrypted data using the decryption key produced after authentication, keeping important information safe from prying eyes [44]. Insights into modern data protection practices and why some companies are taking more preventative and stringent security measures to protect sensitive data from constantly evolving cyber threats can be gained by discussing the Zero Trust Policy and its relevance to using AES 256 in data security. Organizations looking to improve their data security posture might benefit from advice and best practices gleaned from researching the use of Zero Trust and AES 256 in data protection [45].

2.3.3 *Intrusion Detection Systems (IDS)*

IDS tools monitor network traffic for signs of malicious activity or policy violations. They can be network-based or host-based and are essential for identifying potential cyber threats. Intrusion Detection Systems, sometimes known as IDS, are the watchful gatekeepers of contemporary cybersecurity. They provide essential protection against the ever-changing spectrum of cyber threats. These systems are necessary to protect networks, systems, and data. They provide a proactive method of locating and reacting to possible security breaches, which is quite beneficial. This book looks into intrusion detection systems, examining their many varieties, the underlying concepts of how they operate, and their role in current cybersecurity.

Various Categories of IDS

There are two basic varieties of intrusion detection systems, each of which was developed to address a different kind of security concern:

1. Network-Based Intrusion Detection Systems (NIDS): NIDS are positioned at important places within a network to monitor the traffic moving across the network actively. They examine the network packets, looking for odd patterns, known attack signatures, and other unusual occurrences. If suspicious behavior is discovered, the generation of warnings by NIDS enables security staff to react promptly. NIDS effectively spots threats traversing the network, such as attempts to gain unauthorized access, malware, and denial-of-service assaults [46].
2. Host-Based Intrusion Detection Systems (HIDS) are security measures deployed directly on individual hosts or computers. They concentrate on the actions occurring on the host, such as monitoring system files, attempts to log in, and the running of programs. HIDS are especially efficient in identifying suspicious activity at the host level, such as unauthorized changes of vital system files or the presence of malware. Other examples of host-level suspicious actions include the existence of phishing or spam emails [47].

The Fundamentals of IDS Operation

IDS make use of a variety of methodologies, with two basic approaches serving as the primary ways [27, 28]:

1. IDS employs prepared signatures or patterns of known attacks to detect potential threats using the signature-based detection technique. The IDS generates an alert if the network traffic or system behavior fits one of these signatures. Because signature-based detection is efficient at locating known dangers, it is an essential component of any complete security approach.
2. Anomaly-Based Detection Anomaly-based IDS provide a baseline of typical network or system behavior before beginning their analysis. Any changes from this baseline are regarded as suspicious and will cause alarms to be triggered. This method is especially effective for finding previously new threats or zero-day assaults, which do not have recognized signatures and are thus difficult to detect.

Important Characteristics and Capabilities of the IDS

IDS include several critical features and functions, each of which contributes to the overall efficacy of these systems in preserving network security [26].

1. Real-Time Alerting: IDS can provide real-time notifications whenever suspicious behavior is discovered. These notifications are essential to respond promptly and protect against any dangers.
2. Logging and Reporting: IDS give a historical record of security occurrences by generating logs and reports on detected events. These logs are very helpful for post-incident investigation and reporting on compliance.
3. Active and Passive Response: Some IDS can take active actions to counteract potential dangers, such as banning hostile IP addresses or cutting off suspicious network connections. Others only offer notifications and depend on security professionals to take necessary action.
4. Scalability is a feature of IDS systems that allows them to adapt to the requirements of various network sizes, making them appropriate for both small companies and big corporations.
5. Personalization: An IDS may be tailored to concentrate on certain areas of interest, such as database security, web application security, or network perimeter defense.
6. Integration: Many IDS can integrate with other security technologies, such as firewalls and Security Information and Event Management (SIEM) systems, to improve overall security measures and response capabilities.

Relevance of IDSs in Today's World

There are several reasons why intrusion detection systems are such an important component of contemporary cybersecurity [48, 49]:

1. Early Threat Detection: IDS can give early notice of potentially malicious actions or prospective security breaches. This enables security teams to react before major harm has been caused.
2. Anomaly-based detection may discover zero-day attacks, which offer protection against previously unknown threats that do not have established signatures.
3. Compliance and Reporting: IDS are responsible for generating logs and reports, which help in post-incident analysis and make it easier to comply with regulatory standards.
4. Network Security: Network intrusion detection systems (NIDS) defend the network perimeter, while host intrusion detection systems (HIDS) protect individual hosts and devices, improving the network's overall security.
5. Proactive Security: IDS are proactive, which helps organizations avoid security events before they develop.

Intrusion Detection Systems are the watchful sentinels in the ongoing fight against online dangers. They are an essential component of contemporary cybersecurity because of their capacity to recognize both known and unknown dangers, to provide

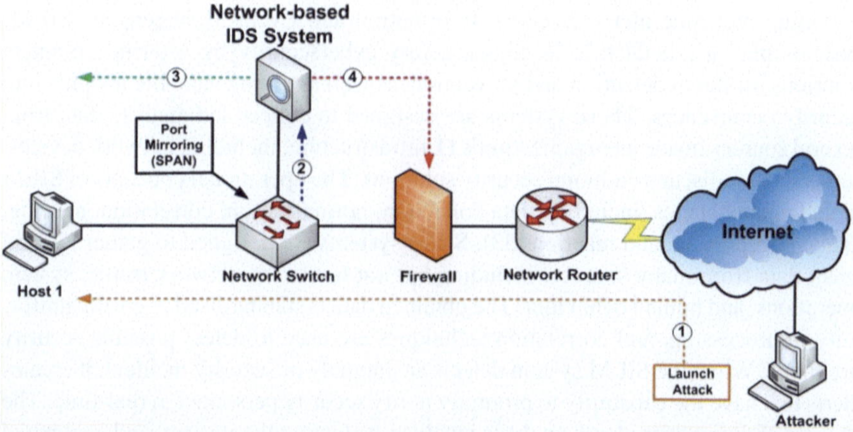

Fig. 2.2 An overview of IDS [52]

warnings in the here and now, and to keep historical records of occurrences. IDS allows organizations to pro-actively protect their networks and systems against a wide variety of security hazards by using a mix of signature-based and anomaly-based detection techniques. Intrusion Detection Systems continue to be vital instruments for protecting sensitive data and maintaining network security, despite the nature of cyber threats is always shifting and developing [50].

It is recommended that a Systematic Literature Review (SLR) on the topic of "Malnutrition Among the Aged Population in Low and Middle-Income Countries" be carried out using various databases and sources. Databases such as PubMed, Google Scholar, WHO, Cochrane Library, ResearchGate, Agricola, Embase, academic journals, UNICEF, NIA, and open access repositories are considered key resources. These sources provide an in-depth analysis of the challenges that low- and middle-income nations face regarding nutrition, aging populations, and public health. The publications and reports of the WHO on global health concerns, such as malnutrition among older people in low and middle-income countries, are also significant sources of information. The National Institute on Ageing (NIA), a division of the National Institutes of Health in the United States, is responsible for conducting and supporting research relating to aging and disorders associated with aging. Open access repositories and the institutional repositories maintained by institutions can store research papers and theses that are relevant to the issue [51]. Thus, Fig. 2.2 illustrates an overview of IDS.

2.3.4 Security Information and Event Management (SIEM)

SIEM solutions collect and analyze data from various sources, including logs and security events. They help in identifying anomalies and potential threats and providing real-time alerts. Security Information and Event Management (SIEM) systems play a crucial role in contemporary cybersecurity by offering complete solutions for threat detection and prevention, as well as giving valuable insights into security occurrences. These systems are designed to collect and analyse data from several sources inside an organization's IT infrastructure, including network devices, servers, firewalls, and endpoint security solutions. The operational concepts of SIEM include many stages, including data collection, normalisation, correlation, alerting, analysis, reporting, and reaction [53]. SIEM systems are designed to gather log and event data from many sources, including but not limited to network traffic, system operations, and human behaviour. The obtained data is standardised to ensure uniformity in processing, and correlation techniques are used to detect possible security problems. When the SIEM system detects an anomaly or security incident, it creates alerts that have the capability to promptly notify security personnel in real-time. The process of doing an in-depth analysis involves evaluating the severity and contextual factors associated with each occurrence, resulting in the creation of comprehensive reports and dashboards that cater to the needs of security professionals and executives [54].

Certain SIEM systems include the capability to seamlessly incorporate with other security technologies, hence enabling the automation of reactions to security events. These responses may include the isolation of compromised endpoints or the prevention of traffic from malicious IP addresses. SIEM systems possess many prominent attributes and functionalities, including log aggregation, alert creation, incident identification and response, compliance reporting, customization capabilities, historical data retention, and seamless interface with diverse security and network devices. SIEM systems are of considerable significance in contemporary cybersecurity due to many factors. These systems provide a comprehensive perspective on the security status of an organisation, facilitating the identification of various types of risks such as malware infections and insider threats. In addition, these systems play a role in promoting adherence to regulatory requirements via the provision of comprehensive reports and audit trails. They effectively consolidate log data from many sources, therefore enhancing efficiency. Furthermore, they contribute to the reduction of dwell time and ensure constant monitoring of the security environment [55]. In summary, it can be concluded that SIEM systems have emerged as vital elements within an organization's cybersecurity framework, offering a complete and proactive approach to safeguarding against cyber threats. SIEM solutions play a crucial role in safeguarding the integrity and confidentiality of organisational data within a more hostile digital environment by consolidating security information and delivering timely cautionary remarks. Figure 2.3 shows the components of SIEM.

2.3.5 Firewalls

Firewalls control incoming and outgoing network traffic based on an organization's established security policies. They are essential for network security and can be

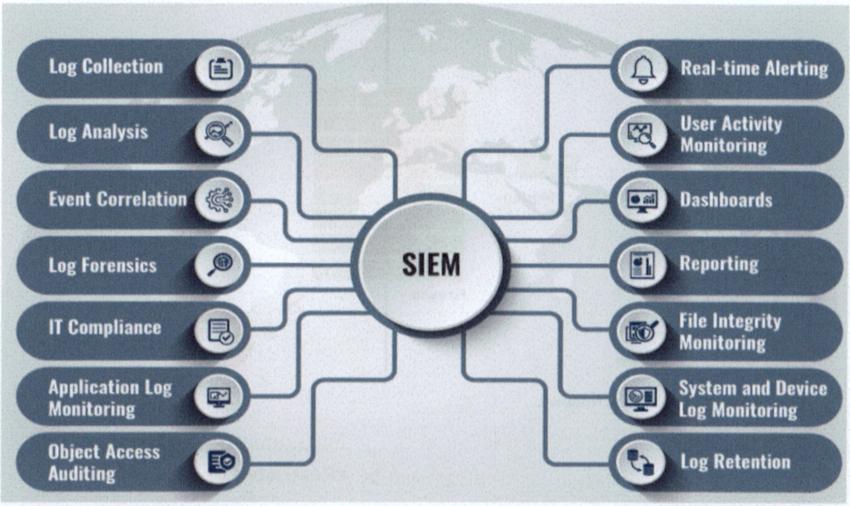

Fig. 2.3 Components of SIEM [56]

implemented at various levels, including the perimeter and within internal network segments. Firewalls against cyber attacks are an essential component in the process of shielding computer networks and systems from a variety of online assaults, including ransomware, malware, phishing attacks, and efforts to break in. They act as the first line of defence, preventing harmful or unauthorised traffic from entering the network and preserving the integrity of the network as a whole. They also play an important part in maintaining regulatory compliance, which helps ensure that corporate activities function smoothly without interruptions brought on by cyberattacks [57].

There are many different kinds of firewalls to protect against cyber attacks, and each one is designed to address a particular vulnerability. Firewalls that use packet filtering perform their functions at the packet level of the network. They examine each data packet and decide whether or not to allow or reject it depending on the criteria that have been specified. Stateful inspection firewalls provide superior security than packet filtering firewalls since they investigate the current status of all active connections. Proxy firewalls function as mediators between requests made on an internal network and those made on an external network [58]. They conceal information pertaining to the internal network while also carrying out extensive content inspection. Application layer firewalls operate at the application layer, and they recognise certain programmes as well as the protocols that are linked with them. Next-generation firewalls, also known as NGFWs, combine the functionality of existing firewalls with that of more modern security measures, such as intrusion prevention, deep packet inspection, and antivirus protection. Access control, inspection of incoming and outgoing traffic, stateful monitoring, intrusion detection and prevention, and content filtering are the primary responsibilities of firewalls designed to protect against cyber-attacks. These types of firewalls regulate access to networks

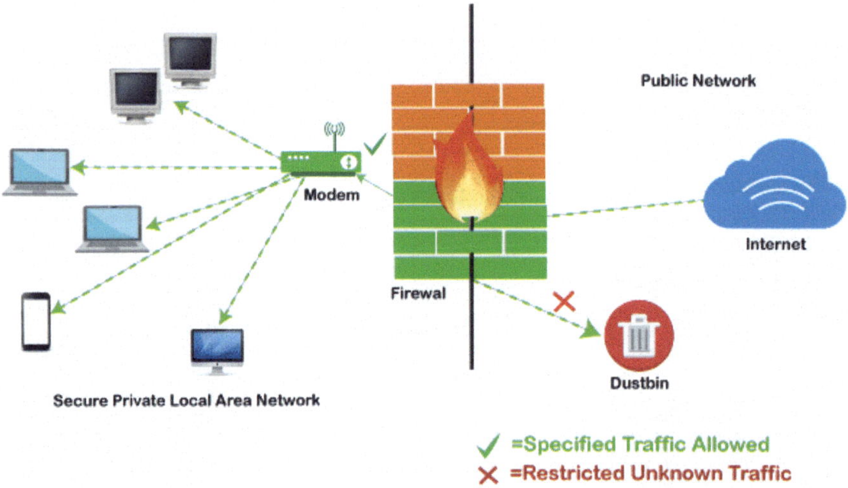

Fig. 2.4 An Overview of cyber-firewalls [60]

and systems by allowing or blocking traffic depending on rules and policies that have been set. In addition to this, they do stateful monitoring, content screening, detection and prevention of intrusions, and traffic inspections. Finally, cyber-attack firewalls are critical for minimising attacks, safeguarding sensitive data, preserving network integrity, and assuring operational continuity. In a world that is becoming more linked, their function will continue to be important in the never-ending struggle to protect digital assets and preserve user faith in an increasingly interdependent environment [59], as illustrated in Fig. 2.4.

2.3.6 Encryption Tools

Encryption protects data both in transit and at rest. Tools for encrypting data ensure that it remains secure and unreadable even if it's compromised, without the proper decryption keys. With the ongoing expansion of the digital ecosystem, ensuring the safeguarding of sensitive information has emerged as a critical need. The use of encryption techniques has become vital in safeguarding against unauthorised intrusion, therefore guaranteeing the confidentiality and integrity of data [61]. This article examines encryption technologies, investigating their importance, features, and the crucial role they fulfil in protecting data in the contemporary digital era.

The Significance of Encryption Tools

The advent of the digital era has brought out a plethora of technical advancements; nevertheless, it has also unveiled weaknesses that malicious actors want to capitalise on. Encryption technologies play a crucial role in many contexts due to their inherent significance [62, 63].

1. Data privacy is safeguarded by the use of encryption methods, which serve to prevent unauthorised individuals from gaining access to sensitive data. Consequently, only those parties who have been granted explicit authorization are able to decrypt and comprehend the information.
2. Security and Confidentiality: The assurance is provided that sensitive data remains secure and unaltered, hence preventing unauthorised access by cybercriminals and those without proper authorization.
3. Compliance with data protection standards is obligatory in several businesses. Encryption plays a crucial role in fulfilling these compliance obligations.
4. Encryption techniques are used to safeguard communication channels, therefore guaranteeing the confidentiality of sensitive information during the process of transmission.
5. Data at rest protection involves the use of encryption measures to ensure the security of data that is kept on various devices, servers, or cloud platforms.

Types of Encryption Tools

Encryption tools are available in a diverse range of formats, each specifically tailored to cater to distinct use scenarios [64, 65].

1. Symmetric encryption is a cryptographic technique that use a single secret key for both the encryption and decryption processes of data. The proposed system demonstrates high efficiency; nonetheless, it necessitates the implementation of robust key management protocols to ensure security.
2. Asymmetric encryption, often known as public-key encryption, employs a dual set of keys, consisting of a public key for the purpose of encryption and a private key for decryption. Secure communication and digital signatures are often used applications.
3. The concept of "End-to-End" refers to a system or process that encompasses the Encryption is a process that guarantees the encryption of data by the sender and its subsequent decryption by the receiver. The protection of user privacy is a regularly used feature in messaging programmes.
4. Full Disc Encryption (FDE) is a software programme designed to encrypt a whole storage device, such as a hard disc, with the purpose of safeguarding all data stored on it.
5. File and folder encryption technologies let users to apply encryption to individual files or folders, hence enhancing the security measures for designated data.

The Primary Roles of Encryption Software

The main purposes of encryption technologies encompass [43, 66]:

1. Data encryption involves the use of algorithms to transform plain text data into a ciphertext that is rendered incomprehensible without the corresponding decryption key.

2. Key management is responsible for ensuring the safe storage and administration of keys in order to mitigate the risk of unauthorised access.
3. Digital signatures are a feature offered by some encryption technologies that enable the creation and verification of digital signatures, therefore establishing the validity and integrity of data.
4. The use of encryption techniques ensures the security of data throughout its transit over networks, therefore safeguarding it against unauthorised interception.
5. Access control refers to the mechanism used to regulate the entry and utilisation of encrypted data by necessitating the provision of authentication and authorization credentials for the purpose of decryption.

In an era characterised by the proliferation of digital technology and the accompanying risks of cyber attacks and data breaches, encryption solutions serve as a robust safeguard for protecting sensitive information. The relevance of these technologies is in their capacity to guarantee the confidentiality and integrity of data, assure compliance with regulations, and provide secure communication. The use of symmetric encryption for data protection at rest, asymmetric encryption for secure communications, and end-to-end encryption for private messaging is crucial in the continuous endeavour to preserve digital assets and uphold user confidence. Encryption technologies are considered indispensable in the contemporary digital environment, as they provide a strong safeguard against the always changing array of cyber risks. By using encryption, data may be kept secret and secure. Encryption tools is illustrated in Fig. 2.5.

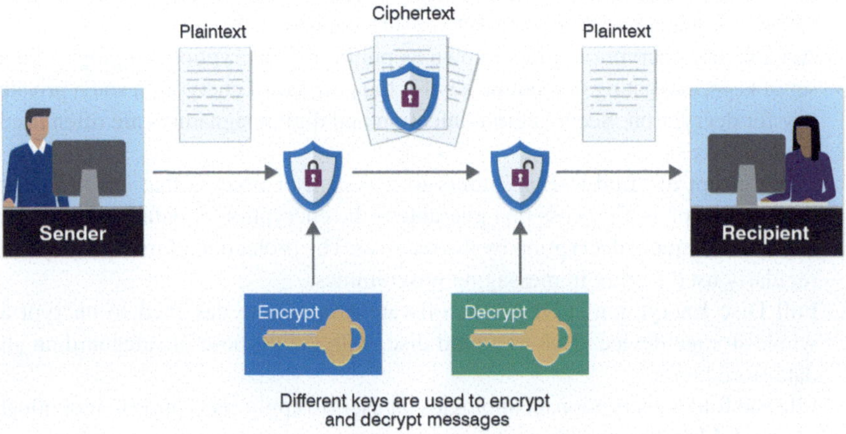

Fig. 2.5 Encryption tools [64]

2.3.7 *Access Control and Identity Management*

These tools control user access to Big Data resources. They include solutions for multi-factor authentication, role-based access control, and Single Sign-On (SSO) systems. Access control and identity management play crucial roles in ensuring the security of information and safeguarding data. Access control is a mechanism that governs the authorization of persons or systems to access certain resources or systems, therefore guaranteeing that only authorized entities are granted access. Physical access control may be used by using various methods such as biometric authentication, access cards, or personal identification numbers (PINs) to ensure the security of physical places. In the field of information technology and data security, the implementation of logical access control serves the purpose of limiting and regulating the entry to computer systems, networks, and data. This is achieved via the use of various approaches such as user authentication, authorization levels, and the application of role-based access control (RBAC).

Role-based access control (RBAC) is a system that streamlines the administration of access control and bolsters security measures. Mandatory access control (MAC) is a security mechanism that imposes access control policies based on security labels. It is often used in government and military contexts to safeguard confidential data. Identity management refers to the systematic procedure of verifying and validating the identities of persons or systems, therefore granting them the necessary permissions and privileges to access various resources and services. The fundamental components of identity management include identification, authentication, authorization, single sign-on (SSO), identity federation, as well as providing and de-provisioning.

Access control and identity management play a crucial role in ensuring security and safeguarding data in various physical and digital environments. Access management plays a crucial role in mitigating the potential risks associated with unauthorized access, data breaches, and security events by ensuring that appropriate individuals or systems are granted access to the necessary resources. Single Sign-On (SSO) facilitates users' access to different systems or apps by using a unified set of login credentials. On the other hand, identity federation empowers users to access resources across several companies or domains by leveraging their pre-existing authentication credentials.

In brief, access control and identity management are essential elements of security and data safeguarding, guaranteeing that exclusively authorized personnel or systems possess the capability to access designated resources or data, as illustrated in Fig. 2.6.

2.3.8 *Security Analytics Tools*

Security analytics technologies play a crucial role in enabling enterprises to efficiently identify and address cybersecurity risks. These technologies use a range of methodologies including data analysis, machine learning, and artificial intelligence to

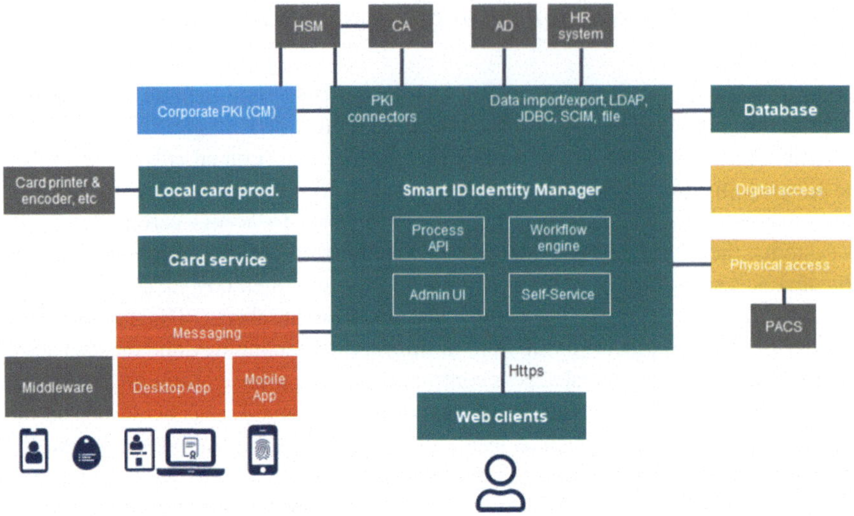

Fig. 2.6 Access control and identity management

detect potentially illicit activity and possible breaches in security. Several important security analytics tools are:

1. Security Information and Event Management (SIEM) systems are designed to gather and analyze data from many sources in order to actively monitor security events, identify deviations from normal patterns, and provide incident response capabilities.
2. User and Entity activity Analytics (UEBA): UEBA technologies are used to observe and analyze the activity of users and entities, establishing standard profiles and issuing notifications when deviations occur, perhaps indicating the presence of insider threats or compromised accounts.
3. Network Traffic Analysis (NTA) refers to the process of examining network traffic patterns in order to detect and identify potentially harmful or unauthorized actions. This includes the identification of anomalous traffic flows or the unauthorized transfer of data outside of the network. NTA tools are specifically designed to facilitate this analysis and aid in the identification of suspicious or malicious behavior.
4. Endpoint Detection and Response (EDR) solutions are designed to actively monitor and promptly react to security risks occurring at the endpoint level. These solutions are capable of identifying and effectively containing many forms of malicious activities, including malware, fileless assaults, and other potential threats that may target individual devices.
5. Vulnerability management solutions include a range of tools that serve the purpose of identifying and prioritizing vulnerabilities inside systems and applications. These tools provide risk assessment capabilities and provide assistance for remedial actions.

6. Threat intelligence platforms include a range of technologies that systematically gather and analyze data from many sources. These platforms serve the purpose of furnishing valuable insights about prevailing threats, attack methodologies, and identified threat actors. By engaging in proactive threat hunting and defense, organizations may use this knowledge to enhance their security posture.
7. The integration of machine learning and artificial intelligence (AI) into security analytics systems enables the detection of sophisticated and dynamic threats via the identification of trends, anomalies, and previously unidentified risks.
8. Tools for Log Analysis: Log analysis tools are used to parse and analyze log files with the purpose of identifying security events, abnormalities, and instances of illegal access.
9. Cloud Security Posture Management (CSPM) refers to a set of solutions that are designed to enhance the security of cloud environments. These solutions primarily revolve on the monitoring of settings and ensuring compliance with established security best practices.
10. Incident Response Platforms (IRP): IRP systems serve to optimize incident response procedures by enhancing communication and cooperation across incident response teams, automating workflow processes, and offering documentation and reporting functionalities.
11. The issue of network and security is of great importance in the field of information technology. Forensic tools play a vital role in the examination of security events and breaches, as they enable the collection and analysis of data pertaining to these occurrences, ultimately facilitating the identification of their extent, consequences, and source.
12. Deception technologies include the use of tools that generate counterfeit assets, then issuing alerts upon the engagement of attackers with these decoys. This strategic approach effectively redirects attackers away from high-value assets, while concurrently facilitating the timely identification of potential threats.
13. Security Orchestration, Automation, and Response (SOAR) refers to a set of platforms that provide the automation of security processes, orchestration of security tools, and response to security events. These platforms aim to enhance efficiency and consistency in incident management.

2.4 Network Segmentation

Fundamentally, the Cisco Identity Service Engine is engineered to implement uniform and expandable access control policies across a comprehensive range of network devices, users, and endpoints. This feature allows organizations to establish detailed access regulations contingent upon user identities, device classifications, geographical locations, and additional contextual variables. Cisco ISE facilitates the implementation of a Zero Trust security framework within organizations, wherein constant monitoring and verification of all network traffic and access attempts occur [67].

One of the primary characteristics of Cisco ISE is its capacity to facilitate a diverse array of authentication mechanisms, encompassing 802.1X, MAC-based authentication, and web authentication. The flexibility provided enables organizations to select the most appropriate authentication method for various user groups and devices, thereby ensuring a user experience that is both seamless and secure [68].

An additional crucial element of Cisco ISE pertains to its capacity for integration. The system can seamlessly integrate with a wide range of network infrastructure elements, including switches, routers, wireless controllers, and firewalls, to uniformly enforce access policies throughout the entirety of the network [69]. The integration of this technology additionally facilitates instantaneous visibility and management of network operations, empowering administrators to promptly identify and address security vulnerabilities. In addition to its core functionalities, Cisco ISE offers advanced profiling and posture assessment capabilities. The system can autonomously detect and categorize interconnected devices by analyzing their distinctive attributes, including their operating system, device classification, and patterns of application utilization. Furthermore, it can evaluate endpoints' security stance, guaranteeing that devices that adhere to compliance standards and possess adequate security are granted entry to the network [70].

There are numerous advantages associated with Cisco ISE. First and foremost, it improves network security by offering a comprehensive view of user and device activities. This technology aids organizations in identifying and thwarting unauthorized access attempts and potential security breaches [71]. Additionally, utilizing a centralized and user-friendly interface streamlines the administration of access policies, resulting in decreased administrative burden and promoting uniform policy implementation. Cisco Identity Services Engine (ISE) holds significant importance within the contemporary business environment, characterized by the prevalence of remote work and the widespread usage of mobile devices, contributing to network security's heightened intricacy [72]. Cisco ISE aids organizations in mitigating the risks associated with the increasing number of cyber threats by implementing access policies contingent upon user identity and device context.

Cisco ISE is pivotal in contemporary network security and access management. The capability to implement detailed access policies that consider user identities, device types, and contextual factors empowers organizations to adopt a zero-trust security framework, thereby protecting their networks against emerging cyber threats. Cisco ISE is an indispensable resource for organizations aiming to improve their network security stance and guarantee a secure and efficient user experience, owing to its sophisticated functionalities, seamless integration capabilities, and ability to provide real-time visibility [73].

2.5 Network Firewalls

2.5.1 Cisco Adaptive Security Appliance (ASA)

The Cisco Adaptive Security Appliance (ASA) is an all-encompassing security solution that offers high-level network protection capabilities to businesses of any size. It is a versatile firewall and security appliance incorporating various security features to defend networks from possible attacks and threats [74]. This section presents an overview of Cisco ASA, including its capabilities, advantages, and role in today's network security landscape. It protects business networks and data centers against various security threats like viruses, malware, unauthorized access, and denial-of-service (DoS) assaults. Because of its flexible functionality, Cisco ASA has developed into an essential component of the network security architecture of many companies [75].

Protection against Firewalls The Cisco ASA performs the function of a firewall by evaluating and filtering traffic to prevent unauthorized access to the network and introducing possible risks. It may block possible attacks or warn administrators about them, preventing security breaches [76]. Visibility and control of apps The Cisco ASA offers granular application visibility and control, which enables administrators to oversee and limit the use of certain apps on the network. Cisco ASA's all-in-one strategy integrates several security features into a single platform to offer complete protection against various threats. Cisco ASA's centralized administration interface enables administrators to create, monitor, and administer security rules from a single console [52].

The Cisco ASA's intrusion prevention system (IPS) and advanced malware protection (AMP) features to assist in identifying and reacting to security threats in real-time, enhancing the enterprise's proactive security posture. The Cisco ASA is an adaptable and potent security appliance that shields contemporary networks from various security risks [52]. Because of its extensive security features, scalability, and administrative simplification, it is a vital component of the network security architecture of enterprises all over the globe. The capacity of Cisco ASA to respond to changing cyber threats and efficiently secure networks assures the continual protection of sensitive data and vital assets [77].

2.5.2 Palo Alto Networks Next-Generation Firewalls

Next-Generation Firewalls (NGFWs) from Palo Alto Networks are cutting-edge security appliances known for their dependable and all-encompassing network protection. They include cutting-edge technology and features beyond conventional firewall capabilities to identify and stop advanced cyber threats [78]. This section investigates the importance of NGFWs in the context of current network security issues. Next-generation firewalls are essential to network security since traditional firewalls are inadequate against modern attacks. To help security teams monitor encrypted traffic for risks, Palo Alto NGFWs provide SSL decryption and inspection. Proactive threat prevention is made possible by real-time updates on new threats provided by the NGFWs' ability to interface with Palo Alto Networks' WildFire, a cloud-based threat intelligence platform.

Palo Alto Networks Next-Generation Firewall PerksImproved Safety: The danger of data breaches and assaults is greatly reduced with the help of Palo Alto Networks NGFWs, which provide full security capabilities to defend against modern threats. NGFWs are scalable and can be adjusted to meet the security requirements of businesses of varying sizes. NGFWs from Palo Alto Networks: Why They Matter With features like advanced threat prevention, application management, and user-based rules, Palo Alto Networks NGFWs are an integral part of today's network security infrastructure. Organizations serious about network security should invest in Palo Alto Networks Next-Generation Firewalls. The NGFWs from Palo Alto Networks defend businesses from cyber attacks with features like advanced threat prevention, application visibility, and user-based rules. Because of their scalability, ease of administration, and compatibility with threat intelligence, they play an important role in protecting today's complex networks. Palo Alto NGFWs continue to be in the vanguard of network protection, keeping businesses ahead of the curve in combating cybercrime.

2.5.3 Fortinet FortiGate

Networks of any size may benefit from Fortinet FortiGate since it is a scalable and adaptable cybersecurity solution. FortiGate is an NGFW that provides superior protection for corporate networks against cyberattacks. This section will examine how Fortinet FortiGate fits into the larger picture of network security today and discuss its most salient features, advantages, and implications. The leading NGFW, Fortinet FortiGate, meets these problems head-on by providing cutting-edge security features and tight interaction with other Fortinet security solutions [79].

Intrusion prevention, antivirus, and anti-malware functionality are only some of FortiGate's comprehensive threat security features. In today's age of telecommuting, this is more important than ever. FortiGate's single management interface gives a consolidated view of the network security architecture, simplifying the configuration and monitoring of security rules. FortiGate provides various security solutions to defend businesses from all types of cyberattacks. FortiGate is a flexible solution for businesses of all sizes due to its scalability, which allows it to adapt to the varying security requirements of companies of all sizes.

FortiGate may be easily combined with other Fortinet security solutions to provide a unified and effective security infrastructure. Data breaches and cyberattacks may be avoided thanks to FortiGate's proactive threat prevention features, which allow for the early identification and elimination of developing threats. Fortinet FortiGate is crucial to current network security because of the sophisticated threat prevention, application management, and VPN features it provides. Its scalability and robust security features make it a good fit for organizations of all sizes and fields.

The unified management console greatly minimizes the administrative burden associated with security policy setup and monitoring. To sum up, Fortinet FortiGate is an effective and adaptable cybersecurity system that equips businesses with defenses against sophisticated cyberattacks. FortiGate is invaluable for defending vital resources because of its robust security features, scalability, and speed. FortiGate is widely considered the gold standard for network security due to its ability to proactively avoid attacks and its tight integration with the Fortinet Security Fabric.

2.6 Intrusion Detection and Prevention Systems (IDPS)

2.6.1 Snort

Because of the ever-changing nature of cyber threats, organizations must implement effective intrusion detection and prevention systems (IDPS) to keep their networks and data secure. Snort is an open-source intrusion detection and prevention system that has gained widespread popularity. The following sections will examine how Snort has recently improved network security [80]. Introduction to Snort: Snort was created by Sourcefire (now owned by Cisco) and is a free and open-source network intrusion detection and prevention system. Martin Roesch created it in 1998, and since then, it has grown into a prominent IDPS utilized by companies of all sizes. Admins can quickly respond to possible attacks thanks to Snort's real-time network traffic analysis and detection of suspicious or malicious actions [81].

Snort uses signature-based detection, which includes checking network traffic against a massive library of previously identified threat signatures. Snort has two distinct modes: inline, which actively blocks or prevents harmful traffic, and passive, where it only monitors and warns administrators without disrupting network traffic. For businesses on a tighter budget, the open-source nature of Snort makes it a viable option. Because of its open-source status and robust detection capabilities, Snort has become a popular and reliable IDPS among security professionals [82]. Because of its low price and widespread community support, it has become an important tool for companies protecting their networks from cyberattacks.

To sum up, Snort has become an indispensable tool in the fight against cybercrime because of the efficiency and low cost with which it can identify and block network attacks. It protects enterprises' most vital data and infrastructure via signature-based detection, protocol analysis, and real-time alerts. Snort's strong community support and commitment to ongoing development make it an invaluable tool in the battle against cyber threats [83].

2.6.2 Suricata

The Open Information Security Foundation (OISF) created Suricata, an open-source intrusion detection system (IDS), to strengthen network security and defend against cyber attacks. High speed and little effect on network throughput are guaranteed by the multi-threaded nature of this IDS, which handles massive amounts of network traffic over many CPU cores [55]. Suricata's ability to monitor and identify assaults of many varieties is made possible by its compatibility with various network protocols. Suricata uses signature-based detection. The Emerging Threats Open (ETOpen) ruleset, constantly updated to accommodate new and developing threat signatures, allows it to respond to new threats as they appear swiftly. To analyze network traffic patterns and spot complicated attack scenarios, it employs network flow analysis to follow and correlate linked packets [52].

Suricata's strengths lie in its ability to detect in real-time, is scalable, has an open-source community, can be extended, and can integrate with many other security tools and platforms. To defend their networks from increasingly complex cyberattacks, businesses need to implement security solutions that are both reliable and flexible. Suricata is an invaluable complement to any network security architecture because of its multi-threading capabilities, real-time detection, and wide range of supported protocols [84]. Finally, in today's network security world, Suricata is essential for warding against cyberattacks. It is a robust open-source IDS that improves network security and keeps businesses one step ahead of would-be attackers thanks to its multi-threading capabilities, real-time detection, and comprehensive protocol compatibility. Suricata is still a useful tool for cybersecurity experts because of its scalability, flexibility, and integration capabilities, all of which aid in protecting networks and vital data [85].

2.6.3 Snorby

Organizations in the modern cyber environment fight an ongoing war against hackers who want to obtain access to their networks and private information. Network monitoring and intrusion detection are crucial for early identification and mitigation of security issues. Snorby is a free online tool for monitoring and analyzing network intrusions and other security incidents. Snort, a widely used open-source intrusion detection system, is its foundation. Snort analyzes the network traffic and looks for intrusions [48]. Snorby's strong points include its web-based interface, real-time event analysis, adaptable alerts, powerful search and filtering options, and the ability to work with Snort. These capabilities make Snorby a competitive alternative to commercial SIEM systems, as it streamlines security operations, allows instantaneous detection and reaction, and shortens the time needed to remediate attacks [86]. In addition, it has a large and supportive user and development community that is always working to improve it.

Businesses may create a robust security architecture by combining Snort with other open-source security technologies like Snorby. As an open-source platform, it's easily accessible and inexpensive, making it a viable option for businesses of varying sizes and financial means. To sum up, Snorby is a robust open-source network security monitoring and management platform that helps companies improve network security and react quickly and appropriately to possible security issues [87]. Snorby streamlines security operations and equips security analysts with the tools to keep networks and vital data safe with its intuitive web-based interface, real-time event analysis, and configurable alerting rules. Snorby is a cost-effective solution with constant community support and integration capabilities, making it an invaluable tool for cybersecurity experts. It helps to keep networks and organizational assets safe and secure [88].

2.7 Data Loss Prevention (DLP)

A strong approach to cybersecurity must include data loss prevention (DLP). Its purpose is to prevent theft, misuse, and exposure of proprietary information. Symantec DLP and McAfee DLP are two of the most well-known options in the DLP industry. The present work compares and contrasts these two options, diving deep into their strengths and weaknesses in protecting sensitive company information [89]. Symantec Data Loss Prevention is an end-to-end solution for detecting, monitoring and protecting sensitive data across a company's network, endpoints, and cloud infrastructure. McAfee Data Loss Prevention (DLP) is another industry leader in DLP, providing a set of safeguards against both accidental and malicious disclosures of sensitive information. McAfee Data Loss Prevention (DLP) enables data detection, categorization, and monitoring across devices, networks, and clouds. Similarities and differences between Symantec DLP and McAfee DLP Both Symantec DLP and McAfee DLP are feature-rich systems that provide strong data security [90].

There are, however, important distinctions between the two: Symantec Data Loss Prevention may be easily combined with other Symantec cybersecurity offerings. McAfee DLP, on the other hand, is tightly integrated with the rest of McAfee's security package, which improves the company's ability to identify and respond to threats more broadly. Symantec Data Loss Prevention may be deployed on-premises or in the cloud, meeting the needs of businesses with varying IT setups. Regarding reporting and analytics, both options cover all the bases. However, depending on the unique needs and preferences of the company, the user interface and simplicity of use may vary. The performance and scalability of the two options may differ depending on the organization's data environment and its size and complexity [91].

Symantec DLP and McAfee DLP are robust tools for securing and guarding against the loss of private information. Organizations may adopt effective data protection measures and stay in compliance with data privacy standards with the help of their data discovery, categorization, and policy enforcement skills. Organizational goals, current and planned cybersecurity measures, and expansion strategies all factor into the decision between the two options. Both technologies help businesses avoid the constantly shifting cybersecurity environment by protecting sensitive information and intellectual property [92].

2.8 Big Data Backup and Recovery

Business choices and insights in today's data-driven world increasingly depend on big data platforms. However, this growing dependence on big data exposes businesses to new dangers, such as data loss and system outages. Strong data security and disaster recovery solutions are required to lessen the impact of these dangers. Backup and recovery capabilities for large data workloads are provided by many industry-leading solutions, including Cloudera Backup and Disaster Recovery (BDR) and Druva inSync [93]. With features like incremental backups, data replication, granular recovery, and scalability, Cloudera BDR is an ideal choice for Hadoop-based large data systems. Data availability and disaster recovery may be ensured because of its data replication capabilities across clusters and data centers. With Cloudera BDR, businesses may restore individual files or parts of a dataset without affecting the rest.

Druva inSync provides automatic and scalable backup solutions for large data systems through the cloud. It uses the cloud to store and safeguard data, making it both scalable and inexpensive. The solution's backup and recovery processes are automated, providing constant security without human involvement. Data security is assured for the foreseeable future with either Cloudera BDR or Druva inSync, which are scalable systems that can keep up with expanding data volumes. Businesses must thoroughly evaluate their data security requirements, current infrastructure, and financial restrictions to choose the optimal solution for big data workloads. Both options considerably help guarantee data integrity and availability, essential for businesses to ensure continuous operations and reduce risks associated with sensitive information [94].

Organizations have increasing hurdles in managing and securing their rising amounts of data in today's data-driven environment. Two significant data management providers, Commvault and Rubrik, provide complete solutions to these difficulties. Commvault is a comprehensive data management platform that provides a single view of data, allowing effective administration and security across varied settings such as on-premises, cloud, and hybrid configurations. It provides a consolidated data management and protection approach, strong backup and recovery capabilities, archiving and data governance, and easy connectivity with major cloud providers [95].

Rubrik is a cloud data management platform that makes backup, recovery, and data management chores as simple as possible. Its simple user interface, rapid recovery, policy-based automation, and cloud-native data protection simplify securing and managing data across many clouds. Both technologies have cloud integration, although Commvault's is more comprehensive. Another important consideration when deciding between Commvault and Rubrik is scalability. Both systems are scalable and capable of meeting the data security demands of businesses of varying sizes. However, Commvault's greater feature set makes it more appropriate for big corporations with complex data environments. The decision between Commvault and Rubrik will be influenced by an organization's unique data management needs, current infrastructure, and future expansion plans [96].

2.9 Security Information and Event Management (SIEM)

In today's ever-evolving digital environment, businesses are confronted with an ever-increasing number of cyber threats, which implement reliable security information and event management (SIEM) systems necessary to protect vital assets. Leading security information and event management (SIEM) products, including Splunk, IBM QRadar, AlienVault USM, and McAfee Enterprise Security Manager, provide sophisticated capabilities to identify and react to cybersecurity problems. Splunk excels in processing and displaying real-time data, providing a user-friendly interface, customized dashboards, and scalability. IBM QRadar is an enterprise-grade security information and event management (SIEM) system with superior threat detection and incident response capabilities. IBM QRadar uses advanced analytics and AI-driven insights to identify and prioritize possible security risks [97].

SIEM stands for security information and event management, and AlienVault USM is a complete solution that integrates several security capabilities into a single platform. These features include threat intelligence, vulnerability assessment, intrusion detection, and SIEM. Unified security administration, integration of threat information, automatic threat detection, and cost-effectiveness are some of the key characteristics of AlienVault USM. McAfee Enterprise Security Manager (ESM) is an advanced SIEM platform that provides real-time visibility and actionable insights into an organization's security posture. It does this via the use of a centralized dashboard. Real-time monitoring, enhanced correlation, connection with other McAfee products, scalability, and performance are some of the key features that come standard with McAfee Enterprise Security Management [98].

In conclusion, Splunk, IBM QRadar, AlienVault USM, and McAfee Enterprise Security Manager are all powerful security information and event management (SIEM) systems that provide useful capabilities for businesses that want to improve their cybersecurity posture. Splunk is notable for its real-time data processing and user-friendly interface. At the same time, IBM QRadar is a formidable enterprise-grade solution due to its sophisticated threat detection and incident response capabilities. AlienVault USM is an option that is both cost-effective and offers a unified security management strategy, making it a good alternative for small to medium-sized businesses. Real-time monitoring and correlation are two areas in which McAfee ESM shines, making it a good option for big companies that must fulfill complicated security needs. The particular requirements, financial constraints, and pre-existing security architecture of a business will determine the selection of one of these SIEM systems [99].

2.10 Distributed Denial of Service (DDoS) Protection

The increasing sophistication and pervasiveness of cyber threats in the modern digital age have made it imperative for businesses to invest in effective cybersecurity measures. Regarding protecting against distributed denial of service (DDoS) attacks and other forms of cybercrime, two of the most trusted names in the industry are Cloudflare and Arbor Networks. Using a worldwide network to cache material closer to end users and reduce DDoS assaults, Cloudflare offers DDoS protection. Service availability and performance are maintained even during peak use or cyber assaults due to the network's scalability and redundancy. DDoS prevention and network visibility are Arbor Networks' bread and butter [100]. It offers comprehensive threat information and analytics to aid businesses in identifying and mitigating cyber risks. Arbor Networks excels in many key areas: distributed denial of service (DDoS) protection, threat intelligence, network visibility, and sophisticated analytics. Traffic is distributed, and latency is lowered thanks to Cloudflare's content delivery network (CDN) and reverse proxy services. At the same time, real-time threat information is gathered by Arbor Networks' worldwide network of sensors [101].

Both businesses provide easily scalable systems to accommodate increasing traffic and sophisticated threats. The hybrid strategy Arbor Networks uses combines on-premise and cloud-based solutions for scalability, while Cloudflare's dispersed network guarantees the availability and performance of services. Using its worldwide network of sensors, Arbor Networks can deliver real-time insights about developing threats, which is the company's greatest strength. Cloudflare's SIEM (Security Information and Event Management) integration provides additional threat information. Cloudflare and Arbor Networks are top-tier cybersecurity solutions providers; however, Cloudflare is most known for its online security and performance services, while Arbor Networks is best known for its DDoS prevention and network visibility. Businesses must accurately analyze their unique situation to choose the most appropriate cybersecurity solution [102]. Investing in a comprehensive and powerful cybersecurity solution is essential to protect digital assets and guarantee company continuity in today's dangerous cyber world.

2.10.1 Antivirus Software

See Table 2.1.

Table 2.1 Selected review papers

Papers	Concepts		
	A	B	C
[103]	x	x	
[104]	x		
[5]	x		
[105]	x		
[106]			x
[107]		x	
[40]		x	
[108]	x		
[30]	x		
[63]	x		
[95]		x	
[109]		x	
[110]		x	
[111]		x	
[112]	x		
[7]	x		
[74]	x		
[113]	x		
[114]	x		
[115]	x	x	
[116]	x		
[117]	x		
[118]	x		
[119]	x		
[120]	x		
[121]	x	x	
[122]	x		
[123]	x		

2.11 Research Gaps

The literature review reveals that big data analytics is widely acknowledged as a powerful and pertinent technology in cyber-attack detection and prevention. However, it is worth noting that there exists a dearth of comprehensive documentation regarding the actual efficacy of this technology. Moreover, a comprehensive theoretical framework outlining the effective implementation of big data analytics for cyber-attack detection is currently lacking. Such a framework would greatly benefit emerging organizations adopting this technology by minimizing the need for trial-and-error approaches.

References

1. Sarma, S.L.V.V.D., Sekhar, D.V., Murali, G.: Stock market analysis with the usage of machine learning and deep learning algorithms. Bull. Electr. Eng. Inform. **12**(1), 552–560 (2023). https://doi.org/10.11591/eei.v12i1.4305
2. Cremer, F., et al.: Cyber risk and cybersecurity: a systematic review of data availability. Geneva Pap. Risk Insur. Issues Pract. **47**(3), 698–736 (2022). https://doi.org/10.1057/s41288-022-00266-6
3. Kumar, A.A.D., Kusonthammarat, P., Guzman, A.L., Zohuri, B.: Supply chain driven supply and demand augmenting resiliency integrated artificial intelligence. J. Econ. Manag. Res. **3**(1), 1–4 (2022). https://doi.org/10.47363/jesmr/2022(3)146
4. Meneses Silva, C.V., Silva Fontes, R., Colaço Júnior, M.: Intelligent fake news detection: a systematic mapping. J. Appl. Secur. Res.16, no. 2, pp. 168–189, 2021, https://doi.org/10.1080/19361610.2020.1761224.
5. Sun, Z., Wang, Q., Chen, L., Hu, C.: Unmanned technology-based civil-military intelligent logistics system: from construction to integration. J. Beijing Inst. Technol. **31**(2), 140–151 (2022). https://doi.org/10.15918/j.jbit1004-0579.2022.010
6. Tufail, S., Parvez, I., Batool, S., Sarwat, A.: A survey on cybersecurity challenges, detection, and mitigation techniques for the smart grid. Energies **14**(18), 1–22 (2021). https://doi.org/10.3390/en14185894
7. Abdullahi, M., et al.: Detecting cybersecurity attacks in internet of things using artificial intelligence methods: a systematic literature review. Electron **11**(2), 1–27 (2022). https://doi.org/10.3390/electronics11020198
8. Dong, T., Li, S., Qiu, H., Lu, J.: An Interpretable Federated Learning-Based Network Intrusion Detection Framework, pp. 1–12 (2022). http://arxiv.org/abs/2201.03134
9. Zimba, A.: A Bayesian attack-network modeling approach to mitigating malware-based banking cyberattacks. Int. J. Comput. Netw. Inf. Secur. **14**(1), 25–39 (2022). https://doi.org/10.5815/ijcnis.2022.01.03
10. Aziz, S., et al.: Anomaly detection in the internet of vehicular networks using explainable neural networks (xNN). Mathematics **10**(8), 1–23 (2022). https://doi.org/10.3390/math10081267
11. Seungjin, L., Abdullah, A., Jhanjhi, N.Z.: A review on honeypot-based botnet detection models for smart factory. Int. J. Adv. Comput. Sci. Appl. **11**(6), 418–435 (2020). https://doi.org/10.14569/IJACSA.2020.0110654
12. Johri, A., Kumar, S.: Exploring customer awareness towards their cyber security in the Kingdom of Saudi Arabia: a study in the era of banking digital transformation. Hum. Behav. Emerg. Technol. **2023** (2023). https://doi.org/10.1155/2023/2103442
13. Alahmadi, A.A., et al.: DDoS attack detection in IoT-based networks using machine learning models: a survey and research directions. Electron **12**(14), 1–24 (2023). https://doi.org/10.3390/electronics12143103
14. Bures, M., Blazek, P., Nema, J., Schvach, H.: Factors impacting resilience of internet of things systems in critical infrastructure. IEEE Access **21**(5), 2–6 (2022). https://doi.org/10.48550/arXiv.2205.13576
15. Qartah, A.A.: Evolving ransomware attacks on healthcare providers. In: ICASSP—IEEE International Conference on Acoustics, Speech, and Signal Processing, vol. 01, no. August, pp. 12–24 (2020). https://doi.org/10.13140/RG.2.2.23202.45765.
16. Di Maria, E., De Marchi, V., Galeazzo, A.: Industry 4.0 technologies and circular economy: the mediating role of supply chain integration. Bus. Strateg. Environ. **31**(2), 619–632 (2022). https://doi.org/10.1002/bse.2940

17. St-Hilaire, C., Brunila, M., Wachsmuth, D.: High rises and housing stress: a spatial big data analysis of rental housing financialization. J. Am. Plan. Assoc. 1–15 (2023). https://doi.org/10.1080/01944363.2022.2126382
18. Mohammed, M.A., Akawee, M.M., Saleh, Z.H., Hasan, R.A., Ali, A.H., Sutikno, T.: The effectiveness of big data classification control based on principal component analysis. Bull. Electr. Eng. Informatics **12**(1), 427–434 (2023). https://doi.org/10.11591/eei.v12i1.4405
19. Nesi, P., Pantaleo, G.: IoT-enabled smart cities : a review of concepts , frameworks and key technologies. Appl. Sci. (2022)
20. Ji, Y.: Exploratory research on the relationship between digital service supply chain capability and supply chain performance. EUrASEANs J. Glob. Socio Econ. Dyn. **1**(1(32)), 7–20 (2022). https://doi.org/10.35678/2539-5645.1(32).2022.7-20
21. Rusch, M., Schöggl, J.P., Baumgartner, R.J.: Application of digital technologies for sustainable product management in a circular economy: a review. Bus. Strateg. Environ. (March), 1–16 (2022). https://doi.org/10.1002/bse.3099
22. Al-Garadi, M.A., Mohamed, A., Al-Ali, A.K., Du, X., Ali, I., Guizani, M.: A survey of machine and deep learning methods for internet of things (IoT) security. IEEE Commun. Surv. Tutorials **22**(3), 1646–1685 (2020). https://doi.org/10.1109/COMST.2020.2988293
23. Orenga-Roglá, S., Chalmeta, R.: Methodology for the implementation of knowledge management systems 2.0: a case study in an oil and gas company. J. Innov. Knowl. **5**(2), 1–16 (2019). https://doi.org/10.1007/s12599-017-0513-1
24. Kenett, R.S., Bortman, J.: The digital twin in Industry 4.0: a wide-angle perspective. Qual. Reliab. Eng. Int. **38**(3), 1357–1366 (2022). https://doi.org/10.1002/qre.2948
25. Ahsan, T., et al.: IoT devices, user authentication, and data management in a secure, validated manner through the blockchain system. Wirel. Commun. Mob. Comput. **2022** (2022). https://doi.org/10.1155/2022/8570064
26. Gilani, K., Ghaffari, F., Bertin, E., Crespi, N.: Self-sovereign identity management framework using smart contracts. In: *Proceedings of IEEE/IFIP Network and Service Management in the Era of Cloudification, Softwarization and Artificial Intelligence. NOMS 2022* (2022). https://doi.org/10.1109/NOMS54207.2022.9789831
27. Hameed, S., et al.: A scalable key and trust management solution for IoT sensors using SDN and blockchain technology. IEEE Sens. J. **21**(6), 8716–8733 (2021). https://doi.org/10.1109/JSEN.2021.3052009
28. Chehab, M.I., Abdallah, A.E.: Architectures for identity management. In: 2009 International Conference for Internet Technology and Secured Transactions (ICITST), pp. 1–8 (2016). 978-1-4244-5647-5
29. Srivastava, P., Pande, S.S.: A novel architecture for identity management system using virtual appliance technology. In: 2014 Seventh International Conference on Contemporary Computing, pp. 171–175 (2014). https://doi.org/10.1109/IC3.2014.6897168
30. Bhattacharya, S., Kumar, P., Maddikunta, R., Pham, Q.: Deep learning and medical image processing for coronavirus (COVID-19) pandemic: a survey. Sustain. Cities Soc. **65**(November), 102589 (2021). https://doi.org/10.1016/j.scs.2020.102589
31. Yildizbasi, A., Arioz, Y.: Green supplier selection in new era for sustainability: a novel method for integrating big data analytics and a hybrid fuzzy multi-criteria decision making. Soft. Comput. **26**(1), 253–270 (2022). https://doi.org/10.1007/s00500-021-06477-8
32. Hunter, W.C.: J. Smart Tour. **1**(2), 27–36 (2021). http://smarttourism.khu.ac.kr/file/202103/1622686933.pdf
33. Ahmad, M., Amin, M.B., Hussain, S., Kang, B.H., Cheong, T., Lee, S.: Health Fog: a novel framework for health and wellness applications. J. Supercomput. **72**(10), 3677–3695 (2016). https://doi.org/10.1007/s11227-016-1634-x
34. Tafti, A.P., et al.: Adverse drug event discovery using biomedical literature: a big data neural network adventure. JMIR Med. Inf. **5**(4), 1–18 (2017). https://doi.org/10.2196/medinform.9170
35. Schmidt, A.F., Finan, C.: Linear regression and the normality assumption. J. Clin. Epidemiol. **98**, 146–151 (2018). https://doi.org/10.1016/j.jclinepi.2017.12.006

36. Ahsan, M.A.M., et al.: Searching on encrypted E-data using random searchable encryption (RanSCrypt) scheme. Symmetry **10**(5), 1–22 (2018). https://doi.org/10.3390/sym10050161
37. Chen, Q., Bridges, R.A.: Automated behavioral analysis of malware: a case study of wannacry ransomware. In: Proceedings of 16th IEEE International Conference on Machine Learning and Applications ICMLA 2017, vol. 2017-Decem, pp. 454–460 (2017). https://doi.org/10.1109/ICMLA.2017.0-119
38. Jiang, Y., Zhu, Y., Wu, W., Li, D.: Makespan minimization for MapReduce systems with different servers. Future Gener. Comput. Syst. **67**, 13–21 (2017). https://doi.org/10.1016/j.future.2016.07.012
39. Zhang, Z., Wen, F., Sun, Z., Guo, X., He, T., Lee, C.: Artificial intelligence-enabled sensing technologies in the 5G/internet of things era: from virtual reality/augmented reality to the digital twin. Adv. Intell. Syst. **4**(7), 2100228 (2022). https://doi.org/10.1002/aisy.202100228
40. Regona, M., Yigitcanlar, T., Xia, B., Li, R.Y.M.: Opportunities and adoption challenges of AI in the construction industry: a PRISMA review. J. Open Innov. Technol. Mark. Complex. **8**(1) (2022). https://doi.org/10.3390/joitmc8010045
41. Mishra, N., Pandya, S.: Internet of things applicaations, security challenges, attacks, intrusion detection, and future visions: a systematic review. IEEE Access **9**, 59353–59377 (2021). https://doi.org/10.1109/ACCESS.2021.3073408
42. Sahi, S.K.: A study of wannacry ransomware attack. Int. J. Eng. Res. Comput. Sci. Eng. **4**(9), 5–7 (2017)
43. Tsiknas, K., Taketzis, D., Demertzis, K., Skianis, C.: Cyber threats to industrial IoT: a survey on attacks and countermeasures. Preprints (February), pp. 1–26 (2021). https://doi.org/10.20944/preprints202102.0148.v1
44. Masmali, H.H., Miah, S.J.: Emergent insight of the cyber security management for Saudi Arabian universities: a content analysis. Lect. Notes Netw. Syst. **448**, 153–171 (2023). https://doi.org/10.1007/978-981-19-1610-6_14
45. Albalawi, A.M., Almaiah, M.A.: Assessing and reviewing of cyber-security threats, attacks, mitigation techniques in IoT environment. J. Theor. Appl. Inf. Technol. **100**(9), 2988–3011 (2022)
46. Shakhov, V., Koo, I.: Graph-based technique for survivability assessment and optimization of IoT applications. Int. J. Softw. Tools Technol. Transf. **23**(1), 105–114 (2021). https://doi.org/10.1007/s10009-020-00594-9
47. Coulter, R., Pan, L.: Intelligent agents defending for an IoT world: a review. Comput. Secur. **73**, 439–458 (2018). https://doi.org/10.1016/j.cose.2017.11.014
48. Hampton, N., Baig, Z., Zeadally, S.: Ransomware behavioural analysis on windows platforms. J. Inf. Secur. Appl. **40**, 44–51 (2018). https://doi.org/10.1016/j.jisa.2018.02.008
49. Panoff, M., Dutta, R.G., Hu, Y., Yang, K., Jin, Y.: On sensor security in the era of IoT and CPS. SN Comput. Sci. **2**(1), 1–14 (2021). https://doi.org/10.1007/s42979-020-00423-5
50. Orabi, M., Mouheb, D., Al Aghbari, Z., Kamel, I.: Detection of bots in social media: a systematic review. Inf. Process. Manag. **57**(4) (2020). https://doi.org/10.1016/j.ipm.2020.102250
51. Safarov, F., Basak, M., Nasimov, R., Abdusalomov, A., Cho, Y.I.: Explainable lightweight block attention module framework for network-based IoT attack detection. Future Internet **15**(9), 297 (2023). https://doi.org/10.3390/fi15090297
52. Kwon, S., Park, S., Cho, H.J., Park, Y., Kim, D., Yim, K.: Towards 5G-based IoT security analysis against Vo5G eavesdropping. Computing **103**(3), 425–447 (2021). https://doi.org/10.1007/s00607-020-00855-0
53. Sokkalingam, S., Ramakrishnan, R.: An intelligent intrusion detection system for distributed denial of service attacks: a support vector machine with hybrid optimization algorithm based approach. Concurr. Comput. Pract. Exp. **34**(27) (2022). https://doi.org/10.1002/cpe.7334
54. Qiu, Z., Miller, D.J., Kesidis, G.: Flow based botnet detection through semi-supervised active learning. In: ICASSP, IEEE International Conference on Acoustics, Speech and Signal Processing—Proceedings, pp. 2387–2391 (2017). https://doi.org/10.1109/ICASSP.2017.7952584

55. Dehkordi, I.F., Manochehri, K., Aghazarian, V.: Internet of things (IoT) intrusion detection by machine learning (ML): a review **12**(1), 13–38 (2023)
56. Liu, L., Wang, P., Lin, J.: ConFlow: Contrast Network Flow Improving Class-Imbalanced Learning in Network Intrusion Detection, pp. 0–21 (2022)
57. Rose, K., Eldridge, S., Chapin, L.: The internet of things: an overview. Understanding the issues and challenges of a more connected world. Internet Soc. (October), 80 (2015). https://doi.org/10.5480/1536-5026-34.1.63
58. Kumar, A., Lim, T.J.: EDIMA: early detection of IoT malware network activity using machine learning techniques. In: IEEE 5th World Forum on Internet of Things, WF-IoT 2019—Conference Proceedings, pp. 289–294 (2019). https://doi.org/10.1109/WF-IoT.2019.8767194
59. Yoon, J.: Deep-learning approach to attack handling of IoT devices using IoT-enabled network services. Internet Things **11**, 100241 (2020), https://doi.org/10.1016/j.iot.2020.100241
60. Rose, K., Eldridge, S., Chapin, L.: The internet of things: an overview. Understanding the issues and challenges of a more connected world. Internet Soc. **2**(October), 80 (2015). http://electronicdesign.com/communications/internet-things-needs-firewalls-too
61. Hammad, M., et al.: Security framework for network-based manufacturing systems with personalized customization: an industry 4.0 approach. Sensors **23**(17) (2023). https://doi.org/10.3390/s23177555
62. Annu, A., Poriye, M., Kumar, V.: Ransomware: detection and prevention. Int. J. Comput. Sci. Eng. **6**(5), 900–905 (2018). https://doi.org/10.26438/ijcse/v6i5.900905
63. Chandel, S., Cao, W., Sun, Z., Yang, J., Zhang, B., Ni, T.Y.: A Multi-Dimensional Adversary Analysis of RSA and ECC in Blockchain Encryption, vol. 70. Springer International Publishing (2020). https://doi.org/10.1007/978-3-030-12385-7_67
64. Ahmed, A.S., Lawal, M.: A secured framework for short messages service in global system for mobile communication. FUOYE J. Eng. Technol. **7**(2), 133–140 (2022)
65. Wu, X.W., Yang, E.H., Wang, J.: Lightweight security protocols for the internet of things. In: IEEE International Symposium on Personal, Indoor and Mobile Radio Communications, PIMRC, vol. 2017, no. October, pp. 1–7 (2018). https://doi.org/10.1109/PIMRC.2017.8292779
66. Saleem, N., Rahman, A., Rizwan, M., Naseem, S., Ahmad, F.: Enhancing security of android operating system based phones using quantum key distribution. EAI Endors. Trans. Scalable Inf. Syst. **7**(28), 1–8 (2020). https://doi.org/10.4108/eai.13-7-2018.165281
67. Zhang, Y., Nakanishi, R., Sasabe, M., Kasahara, S.: Combining iota and attribute-based encryption for access control in the internet of things. Sensors **21**(15), 1–27 (2021). https://doi.org/10.3390/s21155053
68. Saini, S., Panjwani, D., Saxena, N.: Mobile Mental Health Apps: Alternative Intervention or Intrusion? (2022). http://arxiv.org/abs/2206.10728
69. Tauseef, M., Kounte, M.R., Nalband, A.H., Ahmed, M.R.: Exploring the joint potential of blockchain and AI for securing internet of things. Int. J. Adv. Comput. Sci. Appl. **14**(4), 885–895 (2023). https://doi.org/10.14569/IJACSA.2023.0140498
70. Karoui, K., Ben Ftima, F.: New engineering method for the risk assessment: case study signal jamming of the M-Health networks. Mob. Netw. Appl. (2018). https://doi.org/10.1007/s11036-018-1098-8
71. Li, R., Wang, Q., Wang, Q., Galindo, D.: How Do Smart Contracts Benefit Security Protocols?, pp. 1–29 (2022). http://arxiv.org/abs/2202.08699
72. Bubukayr, M.A.S., Almaiah, M.A.: Cybersecurity concerns in smart-phones and applications: a survey. In: 2021 International Conference on Information Technology, ICIT 2021—Proceedings, no. January, pp. 725–731 (2021). https://doi.org/10.1109/ICIT52682.2021.9491691
73. Azgar, A., Rana, S., Hossain, S., Ferdous, M.J.: Testing challenges for mobile applications: an evaluation and comparative analysis of different testing approaches. Int. J. Res. Innov. Appl. Sci. **07**(04), 07–13 (2022). https://doi.org/10.51584/ijrias.2022.7402

74. Azrour, M., Mabrouki, J., Guezzaz, A., Kanwal, A.: Internet of things security: challenges and key issues. Secur. Commun. Netw. **2021**(May) (2021). https://doi.org/10.1155/2021/5533843
75. Falade, P.V., Ogundele, G.B.: Vulnerability analysis of digital banks. Mob. Appl. **1**(1), 44–55 (2022)
76. Yoro, R.E., Aghware, F.O., Akazue, M.I., Ibor, A.E., Ojugo, A.A.: Evidence of personality traits on phishing attack menace among selected university undergraduates in Nigerian. Int. J. Electr. Comput. Eng. **13**(2), 1943–1953 (2023). https://doi.org/10.11591/ijece.v13i2.pp1943-1953
77. Wang, C., et al.: Accurate sybil attack detection based on fine-grained physical channel information. Sensors **18**(3), 1–23 (2018). https://doi.org/10.3390/s18030878
78. Kishore, N., Senapati, A.: 5G smart antenna for IoT application: a review. Int. J. Commun. Syst. **35**(13), 1–16 (2022). https://doi.org/10.1002/dac.5241
79. Soares, C.B.R.B., et al.: Evaluation of third molar development in the estimation of chronological age. Foren. Sci. Int. **254**, 13–17 (2015). https://doi.org/10.1016/j.forsciint.2015.06.022
80. Alwada'n, T., Al-Tamimi, A.-K., Mohammad, A.H., Salem, M., Muhammad, Y.: Dynamic congestion management system for cloud service broker. Int. J. Electr. Comput. Eng. **13**(1):872 (2023). https://doi.org/10.11591/ijece.v13i1.pp872-883
81. Baraka, H.B., Tianfield, H.: Intrusion detection system for cloud environment. ACM Int. Conf. Proc. Ser. **2014**(September), 399–404 (2014). https://doi.org/10.1145/2659651.2659682
82. Khan, M.A., Salah, K.: IoT security: Review, blockchain solutions, and open challenges. Future Gener. Comput. Syst. **82**, 395–411 (2018). https://doi.org/10.1016/j.future.2017.11.022
83. Sezer, S., et al.: Introduction: what is software-defined networking? Future carrier networks are we ready for SDN? Implementation challenges for software-defined networks background: why SDN? Future Carr. Netw. **51**(7), 36–43 (2013). https://doi.org/10.1109/MCOM.2013.6553676
84. Cui, P., Guin, U.: Countering botnet of things using blockchain-based authenticity framework. In: Proceedings of IEEE Computer Society Annual Symposium on VLSI, ISVLSI, vol. 2019-July, pp. 598–603 (2019). https://doi.org/10.1109/ISVLSI.2019.00112
85. Malgwi, Y.M., Goni, I., Ahmad, B.M.: Artificial neural network model for intrusion detection system. Mediterr. J. Basic Appl. Sci. **06**(01), 20–26 (2022). https://doi.org/10.46382/mjbas.2022.6103
86. Kasinathan, P., Costamagna, G., Khaleel, H., Pastrone, C., Spirito, M.: "DEMO: an IDS framework for internet of things empowered by 6LoWPAN. In: Proceedings of the 2013 ACM SIGSAC Conference on Computer and Communications Security—CCS'13, no. October 2015, pp. 1337–1340 (2013). https://doi.org/10.1145/2508859.2512494
87. Bouacida, N., Mohapatra, P.: Vulnerabilities in federated learning. IEEE Access **9**, 63229–63249 (2021). https://doi.org/10.1109/ACCESS.2021.3075203
88. Al-rimy, B.A.S., Maarof, M.A., Shaid, S.Z.M.: Ransomware threat success factors, taxonomy, and countermeasures: a survey and research directions. Comput. Secur. **74**, 144–166 (2018). https://doi.org/10.1016/j.cose.2018.01.001
89. Batool, S., et al.: Lightweight statistical approach towards TCP SYN flood DDoS attack detection and mitigation in SDN environment. Secur. Commun. Netw. **2022** (2022). https://doi.org/10.1155/2022/2593672
90. Kumar, N., Madhuri, J., Channegowda, M.: Review on security and privacy concerns in internet of things. Turk. J. Physiother. Rehabil. **32**(3), 1–5 (2021). https://doi.org/10.1109/ICIOTA.2017.8073640
91. Al-Shareeda, M.A., Manickam, S., Saare, M.A.: DDoS attacks detection using machine learning and deep learning techniques: analysis and comparison. Bull. Electr. Eng. Inf. **12**(2), 930–939 (2023). https://doi.org/10.11591/eei.v12i2.4466
92. Rajendran, G., Ragul Nivash, R.S., Parthy, P.P., Balamurugan, S.: Modern security threats in the internet of things (IoT): attacks and countermeasures. Proc. Int. Carnahan Conf. Secur. Technol. vol. 2019 (2019). https://doi.org/10.1109/CCST.2019.8888399

93. Da Xu, L., He, W., Li, S.: Internet of things in industries: a survey. IEEE Trans. Ind. Inf. **10**(4), 2233–2243 (2014). https://doi.org/10.1109/TII.2014.2300753
94. Fu, H., Chen, H., Zhu, Y., Yu, W.: FARMS: efficient mapreduce speculation for failure recovery in short jobs. Parallel Comput. **61**, 68–82 (2017). https://doi.org/10.1016/j.parco.2016.10.004
95. Sánchez, M.: A general approach on privacy and its implications in the digital economy. J. Econ. Issues **56**(1), 244–258 (2022). https://doi.org/10.1080/00213624.2022.2025729
96. Çetin, S., De Wolf, C., Bocken, N.: Circular digital built environment: an emerging framework. Sustainability **13**(11), 1–34 (2021). https://doi.org/10.3390/su13116348
97. Jararweh, Y., Al-Ayyoub, M., Darabseh, A., Benkhelifa, E., Vouk, M., Rindos, A.: SDIoT: a software defined based internet of things framework. J. Ambient. Intell. Humaniz. Comput. **6**(4), 453–461 (2015). https://doi.org/10.1007/s12652-015-0290-y
98. Mohammed, B.H., Husairi, A., Sallehudin, H., Alaba, F.A., Safie, N.: A conceptual framework for securing IoT-BIM. In: Proceedings of the 2022 International Conference on Intelligent and Innovative Computing Applications, no. May, pp. 68–71 (2022). https://doi.org/10.1109/AiIC54368.2022.9914592
99. Raza, S., Wallgren, L., Voigt, T.: SVELTE: real-time intrusion detection in the internet of things. Ad Hoc Netw. **11**(8), 2661–2674 (2013). https://doi.org/10.1016/j.adhoc.2013.04.014
100. Zhou, Q., Li, R., Xu, L., Nallanathan, A., Yanga, J., Fu, A.: Sufficient Reasons for A Zero-Day Intrusion Detection Artificial Immune System (2022). http://arxiv.org/abs/2204.02255
101. Arshad, S., Abbaspour, M., Kharrazi, M., Sanatkar, H.: An anomaly-based botnet detection approach for identifying stealthy botnets. In: ICCAIE 2011—2011 IEEE Conference on Computer Applications and Industrial Electronics, no. Iccaie, pp. 564–569 (2011). https://doi.org/10.1109/ICCAIE.2011.6162198
102. Ceron, J.M., Steding-Jessen, K., Hoepers, C., Granville, L.Z., Margi, C.B.: Improving iot botnet investigation using an adaptive network layer. Sensors **19**(3), 1–16 (2019). https://doi.org/10.3390/s19030727
103. Chen, D.Q., Preston, D.S., Swink, M.: How big data analytics affects supply chain decision-making: an empirical analysis. J. Assoc. Inf. Syst. **22**(5), 1224–1244 (2021). https://doi.org/10.17705/1jais.00713
104. Filho, M.G., Monteiro, L., Mota, R.O., Gonella, J.D.S.L., Campos, L.M.S.: The relationship between circular economy, industry 4.0 and supply chain performance: a combined ISM/fuzzy MICMAC approach. Sustainability **14**(5) (2022). https://doi.org/10.3390/su14052772
105. Pérez-Díez, I., Pérez-Moraga, R., López-Cerdán, A., Salinas-Serrano, J.M., de la Iglesia-Vayá, M.: De-identifying Spanish medical texts-named entity recognition applied to radiology reports. J. Biomed. Semant. **12**(1), 1–13 (2021). https://doi.org/10.1186/s13326-021-00236-2
106. Scianna, A., Gaglio, G.F., La Guardia, M.: Structure monitoring with BIM and IoT: the case study of a bridge beam model. ISPRS Int. J. Geo-Inf. **11**(3), 102–113 (2022). https://doi.org/10.3390/ijgi11030173
107. Huang, T., Kou, S., Liu, D., Li, D., Xing, F.: A BIM-GIS-IoT-based system for excavated soil recycling. Buildings **12**(4), 1–17 (2022). https://doi.org/10.3390/buildings12040457
108. Tavana, M., Shaabani, A., Vanani, I.R., Gangadhari, R.K.: A review of digital transformation on supply chain process management using text mining. Processes **10**(5), 1–19 (2022). https://doi.org/10.3390/pr10050842
109. Butun, I., Osterberg, P., Song, H.: Security of the internet of things: vulnerabilities, attacks, and countermeasures. IEEE Commun. Surv. Tutorials **22**(1), 616–644 (2020). https://doi.org/10.1109/COMST.2019.2953364
110. Kayıkcı, Y., Subramanian, N.: Blockchain interoperability issues in supply chain: exploration of mass adoption procedures. Stud. Big Data **98**, 309–328 (2022). https://doi.org/10.1007/978-3-030-87304-2_13
111. Kehayov, M., Holder, L., Koch, V.: Application of artificial intelligence technology in the manufacturing process and purchasing and supply management. Proc. Comput. Sci. **200**(2019), 1209–1217 (2022). https://doi.org/10.1016/j.procs.2022.01.321
112. Ikegwu, A.C., Nweke, H.F., Anikwe, C.V., Alo, U.R., Okonkwo, O.R.: Big data analytics for data-driven industry: a review of data sources, tools, challenges, solutions, and research

directions. Cluster Comput. **25**(5), 3343–3387 (2022). https://doi.org/10.1007/s10586-022-03568-5
113. Simpson, S.V., Nagarajan, G.: An edge based trustworthy environment establishment for internet of things: an approach for smart cities. Wirel. Netw. **2** (2021). https://doi.org/10.1007/s11276-021-02667-2
114. Khader, R., Eleyan, D.: Survey of DoS/DDoS attacks in IoT. Sustain. Eng. Innov. **3**(1), 23–28 (2021). https://doi.org/10.37868/sei.v3i1.124
115. Kemmoe, V.Y., Kwon, Y., Hussain, R., Cho, S., Son, J.:Leveraging smart contracts for secure and asynchronous group key exchange without trusted third party. IEEE Trans. Dependable Secur. Comput. 1–18 (2022). https://doi.org/10.1109/TDSC.2022.3189977
116. Aroosa, Ullah, S.S., Hussain, S., Alroobaea, R., Ali, I.: Securing NDN-based internet of health things through cost-effective signcryption scheme. Wirel. Commun. Mob. Comput. **2021**(2) (2021). https://doi.org/10.1155/2021/5569365
117. Williams, P., Kaylan, I., Daoud, H., Bayoumi, M.: A survey on security in internet of things with a focus on the impact of emerging technologies. Internet Things **19**(7), 10–23 (2022). https://doi.org/10.1016/j.iot.2022.100564
118. Oyekan, J., Hutabarat, W., Turner, C., Tiwari, A., He, H., Gompelman, R.: A knowledge-based cognitive architecture supported by machine learning algorithms for interpretable monitoring of large-scale satellite networks. Sensors **21**(4267), 1–21 (2021)
119. Khan, R., et al.: A hybrid approach for seamless and interoperable communication in the internet of things. IEEE Netw. **35**(6), 202–208 (2021). https://doi.org/10.1109/MNET.011.2000787
120. Band, S.S., et al.: A survey on machine learning and internet of medical things-based approaches for handling COVID-19: meta-analysis. Front. Public Heal. **10**(June) (2022). https://doi.org/10.3389/fpubh.2022.869238
121. Sergi, I., Malagnino, A., Rosito, R.C., Lacasa, V., Corallo, A., Patrono, L.: Integrating BIM and IoT technologies in innovative fire management systems. In: 2020 5th International Conference on Smart and Sustainable Technologies (SpliTech 2020), vol. 12, no. 3, pp. 1–5 (2020). https://doi.org/10.23919/SpliTech49282.2020.9243838
122. Pu, C.: Sybil attack in RPL-based internet of things: analysis and defenses. IEEE Internet Things J. **7**(6), 4937–4949 (2020). https://doi.org/10.1109/JIOT.2020.2971463
123. Catalano, C., Paiano, L., Calabrese, F., Cataldo, M., Mancarella, L., Tommasi, F.: Anomaly detection in smart agriculture systems. Comput. Ind. **143**(December), 103750 (2022). https://doi.org/10.1016/j.compind.2022.103750

Chapter 3
Identity and Access Management with Symmetric Data Encryption and Network Segmentation in Solving Cyber Attacks on Big Data

A systematic approach is used in the study methodology for researching the application of Identity and Access Management (IAM) with Symmetric Data Encryption and network segmentation. This methodology seeks to investigate the efficacy of these three security measures in mitigating the risk of cyber attacks on big data and the obstacles that this combination presents. The objective is to describe how IAM, symmetric encryption, and Network segmentation work together to create a complete security solution for contemporary businesses and organizations in solving the challenges of cyber attacks on their Big data.

3.1 Identity and Access Management (IAM) Implementation Procedure

In today's digital landscape, securing access to data is crucial for organizations. Identity and Access Management (IAM) plays a vital role in this security, enabling organizations to effectively control, manage, and secure digital identities and access rights. This book comprehensively explains IAM, including concepts like digital identity, authentication, authorization, and access control. It also emphasizes the importance of strategic considerations, such as aligning IAM with business objectives, compliance requirements, and risk assessments [1]. The book then discusses the step-by-step IAM implementation process, which includes defining objectives and scope, conducting identity and access assessments, selecting IAM solutions and technologies, developing policies and procedures, designing the IAM architecture, implementing identity provisioning and lifecycle management, implementing authentication mechanisms, authorization, and access control implementation, integration with existing systems, testing and quality assurance, user training and awareness, deployment and monitoring, and ongoing maintenance and governance [2].

F. A. Alaba and A. Rocha, *The Implication of Cyberattacks on Big Data and How to Mitigate the Risk*, Information Systems Engineering and Management 39,
https://doi.org/10.1007/978-3-031-88570-9_3

The book also discusses best practices for IAM implementation, such as user-centric design, scalability, and continuous monitoring. However, it also highlights challenges organizations must navigate during performance, such as resistance to change, overcomplicating IAM, and neglecting user experience. Real-world case studies illustrate the practical applications of IAM implementation, highlighting the challenges faced, solutions adopted, and outcomes achieved. Future trends in IAM include AI-driven authentication, zero-trust security, and the impact of blockchain on identity management. By following the step-by-step process and learning from real-world case studies, organizations can harness the power of IAM to secure their digital identities and data access effectively [3].

Protecting digital identities and maintaining strict control over who may see sensitive data are paramount for businesses in today's hyperconnected, data-driven world. This extensive article explains the complicated procedure of creating and implementing IAM systems. It investigates the underlying concepts of IAM, navigates the strategic considerations, defines the step-by-step design and implementation process, and dives into best practices. Organizations can improve their cybersecurity, safeguard important assets, and set up effective access controls when they thoroughly understand the complexity involved in designing and implementing IAM [4].

It is impossible to exaggerate how important strong IAM is in today's digital age when data is such a valuable commodity and security breaches are becoming an increasingly pressing issue. IAM acts as the basis for ensuring the safety of digital identities, regulating access permissions, and protecting sensitive data [5–7].

- **Acquire a Solid Understanding of the IAM Process:** This section gives a complete overview of IAM by presenting its key principles, such as access control, authentication, authorization, and identity management. It also highlights the critical role that IAM plays in contemporary cybersecurity.
- **Strategic Considerations in the Design and Implementation of the IAM System:** Planning is the first step in creating an efficient IAM deployment. This section addresses important strategic concerns, such as aligning IAM with corporate goals, reviewing compliance needs, and performing risk assessments.
- **The Formation of the IAM Team:** The design and execution of IAM need a team that has a variety of specialized abilities. In this part, the duties and responsibilities of the team members are outlined, and the necessity of cross-functional teamwork is emphasized.
- **The Methodical Approach to the IAM Design and Implementation Process:** The following is a step-by-step guide to the process of building and deploying IAM systems, which serves as the basis of this portion of the book:

    1. Step 1: Define IAM Objectives and Scope
    2. Step 2: Conduct Identity and Access Assessment
    3. Step 3: Select IAM Solutions and Technologies
    4. Step 4: Develop IAM Policies and Procedures
    5. Step 5: Design the IAM Architecture
    6. Step 6: Identity Provisioning and Lifecycle Management
    7. Step 7: Implement Authentication Mechanisms

8. Step 8: Authorization and Access Control Implementation
9. Step 9: Integration with Existing Systems
10. Step 10: Testing and Quality Assurance
11. Step 11: User Training and Awareness
12. Step 12: Deployment and Monitoring
13. Step 13: Ongoing Maintenance and Governance.

- **Guidelines for the Best Practices in IAM Design and Implementation:** This section presents a collection of best practices to boost the effectiveness of IAM design and implementation, drawing from industry standards and real-world experiences. The list of topics includes user-centric design, scalability, and continuous monitoring.
- **Difficulties and Obstacles:** Despite the many advantages of IAM, there are obstacles that businesses must overcome throughout the design and implementation stages. Inadequate user training, disregarding scalability, and underestimating the difficulty of integration are just some of the major problems discussed in this section.
- **Case Studies Based on Real-World Experience:** This section includes case studies of real-world companies and organizations that have successfully implemented IAM systems to show the practical applications of IAM design and implementation. It shines a light on the difficulties they encountered, the solutions they implemented, and the results they obtained.
- **Emerging Patterns in the Field of IAM:** The IAM sector is constantly developing. This section discusses certain new developments, including multi-factor authentication (MFA), biometrics, and the role of identity and access management (IAM) in the Internet of Things (IoT) and cloud computing.

In summary, in this day and age, when digital identities and data access management are of the utmost importance, IAM is a crucial component of organizations' cybersecurity policies. IAM system design and implementation is a difficult but necessary process that needs careful planning, strategic alignment, and adherence to best practices. This process is hard but crucial. Organizations can leverage the potential of IAM to protect their digital identities, effectively manage access, and improve their cybersecurity posture if they follow the step-by-step procedure and learn from real-world case studies.

3.1.1 Symmetric Data Encryption Implementation Procedure

In today's digital environment, ensuring the safety of one's data is of the utmost importance. Protecting sensitive data from unwanted access is accomplished using a basic method known as symmetric data encryption. The Symmetric Data Encryption Implementation Procedure is investigated in great detail throughout this extensive paper, including other related topics. It addresses the fundamental ideas behind symmetric encryption, lays out an implementation approach in step-by-step detail, goes into best practices, examines important problems, and investigates new trends [8]. Organizations may enhance their cybersecurity defences and protect their most important data assets if they are skilled in symmetric data encryption and have mastered the technique. Today, when data breaches and cyberattacks are becoming more commonplace, protecting sensitive information is an essential concern for businesses of all kinds. Symmetric data encryption is a core strategy for protecting sensitive data from being seen by unauthorized parties [9].

- **Acquire a Working Knowledge of Symmetric Data Encryption:** This section thoroughly comprehends symmetric data encryption by elaborating on fundamental ideas such as encryption algorithms, encryption keys, and ciphertext. This sheds light on the significance of symmetric encryption concerning data protection.
- **The Importance of Using Symmetric Codes to Encrypt Data:** Symmetric encryption is required to protect data while it is at rest and in transit. This section examines the relevance of symmetric encryption in securing sensitive information and ensuring compliance with applicable regulations.
- **Fundamentals of Symmetric Key Encryption:** It is essential to have a solid understanding of the fundamentals of symmetric data encryption before beginning the actual implementation. Encryption algorithms, key management, and encryption modes are all topics that are covered in this section.
- **A Method for Implementing Symmetric Data Encryption Step by Step:** This is the most important part of the article since it explains in step-by-step fashion how to put symmetric data encryption into practice:

    1. Step 1: Define Data Classification and Encryption Policy
    2. Step 2: Select Appropriate Encryption Algorithms and Key Lengths
    3. Step 3: Develop Key Management Procedures
    4. Step 4: Encrypt Data at Rest
    5. Step 5: Implement Encryption for Data in Transit
    6. Step 6: Securely Distribute and Store Encryption Keys
    7. Step 7: Regularly Rotate Encryption Keys
    8. Step 8: Establish Access Controls and Authentication Mechanisms
    9. Step 9: Perform Security Audits and Regular Monitoring
    10. Step 10: Incident Response and Recovery Planning

- **Implementing Symmetric Data Encryption:** This section presents a set of recommended practices that, when followed, will increase the likelihood that an implementation of symmetric data encryption will be successful. Secure key storage, encryption key lifecycle management, and encryption for cloud settings are some of the topics that will be addressed.
- **Difficulties and Considerations:** Symmetric encryption is a formidable tool; nonetheless, it does not come without challenges. This section addresses frequent issues, such as the distribution of keys, the influence on performance, and the complexity of complying with regulations.

- **Case Studies Based on Real-World Experience:** This section offers case studies of businesses from the real world that successfully implemented encryption solutions. The purpose of these case studies is to highlight the practical uses of symmetric data encryption. It sheds light on the difficulties they encountered, the solutions they implemented, and the accomplishments they accomplished.
- **New Directions in Symmetric Data Encryption:** The practice of encrypting data is changing. This section investigates developing themes, such as encryption resistance to quantum computing, homomorphic encryption, and encryption in the context of edge computing.

In summary, when data is a highly valued asset and data breaches may have serious repercussions, it is necessary to grasp the art of symmetric data encryption. In order to successfully implement symmetric data encryption, a protocol must be clearly outlined, encryption algorithms must be chosen with caution, key management must be solid, and best practices should be adhered to. Organizations can enhance their cybersecurity defences and safeguard their most precious data assets from cyberattacks by following the step-by-step deployment procedure and remaining updated about new trends.

3.1.2 Network Segmentation Implementation Procedure

Network segmentation has evolved as an essential method for boosting cybersecurity, preserving sensitive data, and ensuring operational resilience in an age characterized by the unrelenting growth of cyber threats. An exhaustive Network Segmentation Implementation Procedure is provided in this extensive article. This procedure covers network segmentation, strategic considerations, a thorough step-by-step implementation approach, best practices, common problems, and emerging trends [10]. Organizations can bolster their cybersecurity defences, isolating essential assets and proactively addressing the ever-changing environment of cybersecurity threats when they have mastered the complexities of network segmentation and can properly segment their networks [11]. Protecting an organization's network is of the utmost importance today when interconnectivity is the defining characteristic of contemporary corporate operations. Network segmentation has evolved into a foundational concept that enables businesses to restrict access, reduce attack surfaces, and effectively react to cyber threats [12, 13].

- **Recognizing the Value of Network Segmentation:** This section offers a thorough comprehension of network segmentation by elaborating on fundamental ideas such as the advantages of segmentation, the many forms of segmentation, and the part that access restrictions and firewalls play in the process.
- **Network Segmentation and Its Importance:** The contemporary cybersecurity practice relies heavily on network segmentation as an essential component. This section examines the value of network segmentation in preserving business continuity, securing sensitive data, and minimizing the number of attack vectors.

- **The Essential Components of Network Division:** It is essential to have a firm grip on the fundamentals of network segmentation before beginning work on the actual implementation. Specifically, the subjects of network architecture design, access restrictions, segmentation rules, and threat detection are discussed in this section.
- **A Method for Deliberately Implementing Network Segmentation:** This essential part of the article provides a step-by-step breakdown, in excruciating detail, of the process of implementing network segmentation:

  1. Step 1: Define Segmentation Objectives and Scope
  2. Step 2: Conduct Network Assessment and Asset Classification
  3. Step 3: Develop Segmentation Policies and Access Controls
  4. Step 4: Design Segmentation Zones and Network Architecture
  5. Step 5: Implement Segmentation Controls and Isolation Mechanisms
  6. Step 6: Monitor and Manage Network Segmentation
  7. Step 7: Incident Response and Remediation Planning

- **Best Practices for Network Segmentation Implementation:** To guarantee that the network segmentation implementation is a success, this section provides a collection of best practices drawn from industry standards and experiences from the real world. Continuous monitoring, least privilege access, and scalable design are some of the topics that will be covered.
- **Challenges and Issues to Take Into Account:** Although network segmentation is a powerful method for enhancing cybersecurity, implementing this tactic is fraught with difficulties that businesses must master. This part outlines typical challenges, such as complexity, scalability, and user experience, and presents ways to address them. Other sections in the document continue the discussion.
- **Case Studies Adapted From the Real World:** This section includes real-world case studies of enterprises that have successfully used segmentation solutions to highlight network segmentation's practical uses. It sheds light on the difficulties they encountered, the solutions they implemented, and the accomplishments they accomplished.
- **New Segmentation Methods for Networks**: The area of network segmentation is active and always undergoing new developments. This section investigates developing themes such as zero-trust architecture, micro-segmentation, and the interaction of segmentation with cloud and IoT settings, among many others.

In summary, network segmentation is crucial in strengthening an organization's defensive posture, particularly in a cybersecurity environment characterized by persistent and sophisticated attacks. Effectively implementing network segmentation requires a well-structured method, a comprehensive grasp of network interdependence, and a commitment to following best practices. Organizations can improve their cybersecurity posture, safeguard sensitive data, and traverse the dynamic obstacles offered by current cyber threats if they follow the step-by-step implementation process and remain aware of evolving trends.

3.2 Research Design

This work examines the integration of IAM, also known as identity and access management, symmetric encryption with network segmentation using three separate research methodologies: exploratory research, descriptive research, and comparative research. The experimental study aims to learn the theoretical underpinnings of IAM, symmetric encryption, and network segmentation by analyzing the current body of literature, security protocols, and encryption algorithms [14]. This study helps create a strong basis for the following stages by finding possible synergies between symmetric encryption, identity access management (IAM), and network segmentation.

This research aims to explore the implications of cyberattacks on big data and develop strategies to mitigate the associated risks. The objectives include assessing the impact of cyberattacks on big data systems, evaluating the effectiveness of IAM in securing big data, analyzing the role of Symmetric Data Encryption in protecting sensitive data within big data environments, investigating the benefits and challenges of implementing Network Segmentation as a cybersecurity strategy for big data, and proposing comprehensive mitigation strategies that incorporate IAM, Symmetric Data Encryption, and Network Segmentation. The research will use a mixed-methods approach, combining qualitative and quantitative data. It will adopt a sequential exploratory design, starting with qualitative data collection and analysis followed by quantitative data collection and analysis. The qualitative phase will involve in-depth interviews with cybersecurity experts and a review of cybersecurity literature and case studies. The quantitative phase will include online surveys with IT professionals and data security specialists. The sampling strategy will involve purposeful sampling of cybersecurity experts and professionals with expertise in big data security for the qualitative phase. Data saturation will be used as the stopping criterion. For the quantitative phase, random sampling will be conducted from a list of IT professionals and data security specialists. A minimum sample size will be determined using a power analysis to ensure statistical significance.

Data analysis will involve thematic analysis for the qualitative phase to identify common themes and patterns. Descriptive statistics and inferential statistical techniques, such as regression analysis, will be used for the quantitative phase. Ethical considerations include obtaining informed consent, protecting participant anonymity and confidentiality, and adhering to ethical guidelines and obtaining necessary approvals. The research timeline is tentatively set for 12 months, with flexibility for adjustments. A budget will be allocated for participant recruitment, data collection, software tools, and data analysis. Limitations and assumptions include potential limitations related to the availability of experts for interviews and the accuracy of self-reported survey data. Validation and reliability will be ensured through rigorous data analysis techniques and triangulation of findings. Data collection instruments will be developed based on established cybersecurity frameworks and best practices. Data management and storage will be done securely using encryption and access controls. Data interpretation and reporting will involve interpreting findings in the context of the research objectives and reporting results through publications and presentations. Overall, this research design aims to provide valuable insights into the implications of cyberattacks on big data and develop effective strategies to mitigate the associated risks.

Research that is descriptive focuses on existing solutions, looking at real-world situations and implementations in a variety of different businesses. This study investigates what makes the proposed tools ideal for solving the problems of cyberattacks on big data. IAM is responsible for controlling user access; symmetric encryption is used to protect data confidentiality, while Network segmentation allows organizations to define and enforce access policies based on various factors such as user identity, device type, location, and more. The comparative study aims to determine the most successful tactics regarding data security, user comfort, and scalability [15].

The entire research has ramifications that extend beyond theoretical comprehension. Descriptive research offers insights into real-world practices, allowing corporations to gain knowledge from other businesses' experiences and difficulties [16]. Research that compares several options provides decision-makers with the knowledge they need to choose the strategy that will be the most successful in their particular setting. Exploratory, descriptive, and comparative research stages unveil theoretical, practical, and strategic components of applying IAM, symmetric encryption, and network segmentation. The insights from these research methodologies guide companies to implement strong security while guaranteeing an effortless user experience [17].

3.3 Research Approach

Secondary data and primary data, which make up the Big data is a wealth of legacy and current information that gives a basis of concern of complicated security issues, namely Confidentiality, Integrity, and Availability. This work examines the implications of cyber attacks on Big data and emphasizes steps involved in the installation, configuration, and application of the selected tools from this rich resource so as to mitigate the risks involved when cyberattacks occur [18]. Secondary data is critical in research, especially when examining complex themes like IAM, symmetric encryption, and network segmentation. It enables academics to access a wide reservoir of knowledge, including scholarly publications, case studies, industry reports, and whitepapers. Secondary data is valuable because it may speed up the research process by giving context, creating a foundation of knowledge, and delivering insights gleaned thorough examination by experts in the subject. A comprehensive literature study is the foundation for investigating secondary data [19]. Researchers dig through academic publications, research papers, and books that delve into the complexities of IAM, symmetric encryption, and network segmentation. This phase supports comprehending these domains' theoretical underpinnings, practical applications, difficulties, and possibilities. Real-world situations and experiences described in case studies provide essential insights. Researchers may examine how firms effectively combine tools to solve challenges they face and the benefits they will obtain. These case studies offer practical expertise that may be used to guide future implementations. Industry studies give a macroscopic picture of the ecosystem, encompassing trends, problems, and developments in IAMt, symmetric encryption, and network segmentation. Gartner, Forrester, and IDC provide studies that aggregate industry knowledge, making them a helpful resource for comprehending the larger picture [20].

While secondary data may provide a wealth of insights, it is necessary to analyze it critically. Researchers should assess the trustworthiness of the sources, the methodology utilized in the studies, and the relevance to the study setting. Synthesizing the material gathered aids in detecting patterns, trends, and gaps in the available research. Secondary data may help bridge the gap between theoretical understanding and real application. Scholars may get a comprehensive picture that influences decision-making in real-world circumstances by combining findings from several sources [19, 21]. Finally, secondary data is critical in deciphering the complexity of IAM, symmetric encryption, and network segmentation. It provides a plethora of knowledge, from theoretical underpinnings to practical implementations and industry perspectives. Researchers may increase knowledge and informed decision-making in the sphere of Big data security by painstakingly gathering, evaluating, and extracting perspectives from secondary data [22].

3.4 Data Analysis

Applied design research focuses on using existing tools to provide solutions to practical problems. It involves proposing an off-the-shelf software as a solution to a specific research problem after research that involves the following steps has been done. The research process involves the following steps (Chat GPT 2023).

- Identifying a research problem or challenge that can be addressed with the use of software.
- Researching and selecting an appropriate off-the-shelf software that aligns with the problem's requirements.
- Evaluating the software's features, capabilities, and usability in the context of the problem.
- Insatlling and configuring the software to suit the specific needs of the problem.
- Testing and validating the software's effectiveness in addressing the identified problem.
- Assessing user satisfaction and feedback regarding the software's usability and impact.

Concerns about data security, reluctance to adapt to change, limited resources, difficult integration, and increased efficiency are all examples of obstacles that might arise. Some advantages are data-driven insights, better customer experience, a competitive edge, stakeholder engagement, pilot testing, new employee training, and constant monitoring [23].

This research method may reveal hidden gems in interview transcripts and usability testing. Researchers may learn more about participants' points of view by identifying recurring negative feedbacks which can be used as a guide for fixes and change implementation. This analysis serves as a valuable foundation for crafting comprehensive reports, developing strategies, and making informed decisions [24].

3.5 Ethical Considerations

Researchers demonstrate prudence and accountability while doing applied design research, since they encounter ethical dilemmas stemming from the use of pre-existing solutions. It is imperative to approach source material with a sense of respect and reverence. Prior to using any source material, it is essential to obtain proper permission and adhere to copyright regulations. Furthermore, it is crucial to maintain accuracy and preserve the context of the original material. Plagiarism must be strictly avoided, and measures should be taken to protect privacy and maintain confidentiality. In order to ensure a balanced perspective, it is important to consider both bias and objectivity. Additionally, when collecting data, ethical considerations should be carefully evaluated. Respect for social and cultural contexts is paramount, and any limitations of the research should be transparently communicated. Misrepresentation of information should be avoided at all costs, and efforts should be made to minimize any potential harm that may arise from the research [25]. The ethical aims of intellectual honesty and respect for original researchers need the correct citation of sources, obtaining permits, and adhering to copyright regulations. Integrity-preserving practices include the act of appropriately acknowledging the contributions of others by providing fair credit and use quotation marks to denote direct quotations. Safeguarding personal privacy and refraining from divulging confidential information without consent are crucial considerations [26].

In the process of doing secondary data analysis and interpretation, it is essential for researchers to demonstrate awareness and transparency about any inherent or potential biases present within the initial data. The use of data obtained by unethical means should be avoided in academic investigations. It is important to acknowledge and openly discuss the limitations associated with secondary data in research reports. Additionally, it is crucial to maintain a high level of social and cultural consciousness throughout the process of interpreting the data. Ensuring the prevention of distortion and the highlighting of limitations in secondary research is crucial for the objective presentation of the study's strengths and weaknesses. It is incumbent upon researchers to mitigate any adverse consequences arising from their study. Researchers may ensure that their secondary research contributes positively to the corpus of knowledge ethically and responsibly by deliberately navigating these aspects [27].

Chapter Summary

This chapter examines how IAM, Symmetric Data Encryption (SDE), and network segmentation complement one another in data security and what kinds of challenges and opportunities this may provide. Symmetric Data Encryption uses a single key to encrypt data to maintain privacy. Identity and Access Management operates as a gatekeeper to manage resource access based on user identities. At the same time, network segmentation will dynamically enforce segmentation rules to control user and device access to different network segments within and outside an organization. IAM, Symmetric Data Encryption, and Network segmentation work together to provide a solid security structure which can withstand cyber attacks and prevent data leaks. Possible advantages include better data security, more reliable user authentication, and easier administration of access permissions. This chapter focuses on encryption algorithms, key management, and user authentication techniques to examine best practices and strategies for integrating IAM, Symmetric Data Encryption, and network segmentation into enterprise systems. Despite their significant security advantages, IAM, Symmetric Data Encryption, and network segmentation need constant attention, upgrades, and changes to handle evolving threats.

References

1. Aswir, Misbah, H.: Smartphones as personal digital archives? Recentering migrant authority as curating and storytelling subjects. Photosynthetica **2**(1), 1–13 (2018). https://doi.org/10.1007/978-3-319-76887-8
2. Han, J., Ha, M., Kim, D.: Practical security analysis for the constrained node networks: focusing on the DTLS protocol. In: Internet Things (IOT), 2015 5th International Conference, pp. 22–29 (2015). https://doi.org/10.1109/IOT.2015.7356544
3. Ebijuwa, A.S., Mabawonku, I.: Computer self-efficacy as a predictor of undergraduates' use of electronic library resources in federal universities in South-west Nigeria. Glob Knowl. Mem. Commun. **68**(4/5), pp. 323–336, 2019, https://doi.org/10.1108/gkmc-10-2018-0083.
4. Chehab, M.I., Abdallah, A.E.: Architectures for identity management. Internet Technol. Secur. Trans Int. Conf. 1–8 (2016) (978-1-4244-5647-5)
5. Calinaud, V., Kokkranikal, J., Gebbels, M.: Career advancement for women in the British hospitality industry: the enabling factors. Work Employ Soc. **35**(4), 677–695 (2021). https://doi.org/10.1177/0950017020967208
6. Ismail, E.D., Said, S.Y., Jalil, M.K.A., Ismail, N.A.A.: Benefits and challenges of heritage building information modelling application in Malaysia. Environ. Proc. J. **6**(SI4), 179–184 (2021). https://doi.org/10.21834/ebpj.v6isi4.2917
7. Wahab Ahmed, A., Muhammad Ahmed, M., Ahmad Khan, O., Ali Shah, M.: A comprehensive analysis on the security threats and their countermeasures of IoT. Int. J. Adv. Comput. Sci. Appl. **8**(7) (2017). https://doi.org/10.14569/IJACSA.2017.080768
8. Wu, J., Dong, M., Ota, K., Liang, L., Zhou, Z.: Securing distributed storage for social internet of things using regenerating code and Blom key agreement. Peer-to-Peer Netw. Appl. **8**(6), 1133–1142 (2014). https://doi.org/10.1007/s12083-014-0286-y
9. Vasudeva, A., Sood, M.: Survey on sybil attack defense mechanisms in wireless ad hoc networks. J. Netw. Comput. Appl. **120**, 78–118 (2018). https://doi.org/10.1016/j.jnca.2018.07.006

10. Pambhar, H., Aghera, K., Tada, N., Residual, Á.Á.: An advanced web-based bilingual domain independent interface to database using machine learning approach **508**(June), 197–204 (2018). https://doi.org/10.1007/978-981-10-2750-5
11. Ohri, K., Kumar, M.: Review on self-supervised image recognition using deep neural networks. Knowl. Based Syst. **224**, 107090 (2021). https://doi.org/10.1016/j.knosys.2021.107090
12. Comelli, A., et al.: Deep learning-based methods for prostate segmentation in magnetic resonance imaging. Appl. Sci. **11**(2), 1–13 (2021). https://doi.org/10.3390/app11020782
13. Hassan, B., et al.: Deep learning based joint segmentation and characterization of multi-class retinal fluid lesions on OCT scans for clinical use in anti-VEGF therapy. Comput. Biol. Med. **136**(July), 104727 (2021). https://doi.org/10.1016/j.compbiomed.2021.104727
14. Olatoye, T.A.: Analysis of modal choice of residents in Lagos State. J. Hum. Ecol. **65**(1–3) (2019). https://doi.org/10.31901/24566608.2019/65.1-3.3157
15. Pereira, C.S., Durao, N., Moreira, F., Veloso, B.: The importance of digital transformation in your business strategy. MDPI **14**(2), 1–26 (2022). https://blog.workana.com/en/entrepreneurship/the-importance-of-digital-transformation-in-your-business-strategy/
16. Osmani, M.: Construction waste minimization in the UK: current pressures for change and approaches. Proc. Soc. Behav. Sci. **40**, 37–40 (2012). https://doi.org/10.1016/j.sbspro.2012.03.158
17. Arayici, Y., Counsell, J., Mahdjoubi, L., Nagy, G., Hawas, S., Dewidar, K.: Heritage building information modelling (2017). https://doi.org/10.4324/9781315628011
18. Oyedokun, O.O.: Green human resource management practices (GHRM) and its effect on sustainable competitive edge in the Nigerian manufacturing industry. Unpublished Master's Thesis. Nijerya, p. 108 (2019). https://esource.dbs.ie/handle/10788/3829
19. Zhou, H., Milani Fard, A., Makanju, A.: The State of Ethereum smart contracts security: vulnerabilities, countermeasures, and tool support. J. Cybersecur. Priv. **2**(2), 358–378 (2022). https://doi.org/10.3390/jcp2020019
20. Al-Doori, J.A.: The impact of supply chain collaboration on performance in automotive industry: empirical evidence. J. Ind. Eng. Manag. **12**(2), 241–253 (2019). https://doi.org/10.3926/jiem.2835
21. Zhang, Z., Zhang, Y., Zhou, T., Pang, Y.: Medical assertion classification in Chinese EMRs using attention enhanced neural network. Math. Biosci. Eng. **16**(4), 1966–1977 (2019). https://doi.org/10.3934/mbe.2019096
22. Wang, X., Kumar, V., Kumari, A., Kuzmin, E.: Impact of digital technology on supply chain efficiency in manufacturing industry. Lect. Notes Inf. Syst. Organ. **54**(June), 347–371 (2022). https://doi.org/10.1007/978-3-030-94617-3_25
23. Hatem, W.A.: Motivation factors for adopting building information modeling (BIM) in Iraq. Eng. Technol. Appl. Sci. Res. **8**(April), 2668–2672 (2018). https://doi.org/10.5281/ZENODO.1257505
24. Liang, Z., et al.: EEGFuseNet: hybrid unsupervised deep feature characterization and fusion for high-dimensional EEG with an application to emotion recognition. IEEE Trans. Neural Syst. Rehabil. Eng. **29**(3), 1913–1925 (2021). https://doi.org/10.1109/TNSRE.2021.3111689
25. Bubukayr, M.A.S., Almaiah, M.A.: Cybersecurity Concerns in Smart-phones and applications: a survey. In: 2021 International Conference on Information Technology ICIT 2021—Proceedings, no. January, pp. 725–731 (2021). https://doi.org/10.1109/ICIT52682.2021.9491691
26. Otuoze, S.H., Hunt, D.V.L., Jefferson, I.: Neural network approach to modelling transport system resilience for major cities: case studies of Lagos and Kano (Nigeria). Sustainability **13**(3), 1–20 (2021). https://doi.org/10.3390/su13031371
27. Mittal, S., Kumar, V.: Study of knowledge management models and their relevance in organizations. Int. J. Knowl. Manag. Stud. **10**(3), 322–335 (2019). https://doi.org/10.1504/IJKMS.2019.101491

Chapter 4
Result Implementation and Discussion

This chapter provides the result implementation and discussion based on Identity and Access Management (IAM), Symmetric Data Encryption (SDE), and network segmentation. It proposes an integrated approach to mitigate the risk of cyberattacks on big data.

4.1 Result Implementation of Cyberattacks on Big Data

This section will discuss the result of implementing cyberattacks on big data.

4.1.1 Identity and Access Management (IAM)

Protecting sensitive data in the digital era is paramount, particularly within Big Data. Cyberattacks targeting Big Data might have significant ramifications, including compromising sensitive information, monetary setbacks, and impairing one's reputation [1–3]. IAM plays a crucial role in reducing these risks via the effective management of user identities, authentication of users, and control of access rights. IAM provides a comprehensive suite of features, including fine-grained access control, user authentication, authorization management, and real-time monitoring. To effectively establish IAM to ensure security in the context of Big Data, businesses are advised to evaluate their existing environment, deploy a comprehensive user identity management system, use role-based access control mechanisms, apply encryption to safeguard sensitive data and perform periodic audits. The advantages of IAM include risk reduction, adherence to data security standards, improved operational effectiveness, and bolstered reputation. Incorporating IAM into data security plans is becoming more essential as the digital world continues to grow [3–5].

F. A. Alaba and A. Rocha, *The Implication of Cyberattacks on Big Data and How to Mitigate the Risk*, Information Systems Engineering and Management 39,
https://doi.org/10.1007/978-3-031-88570-9_4

Figure 4.1 shows the IAM mitigating approach for cyberattacks on big data, and Table 4.1 provides the algorithm.

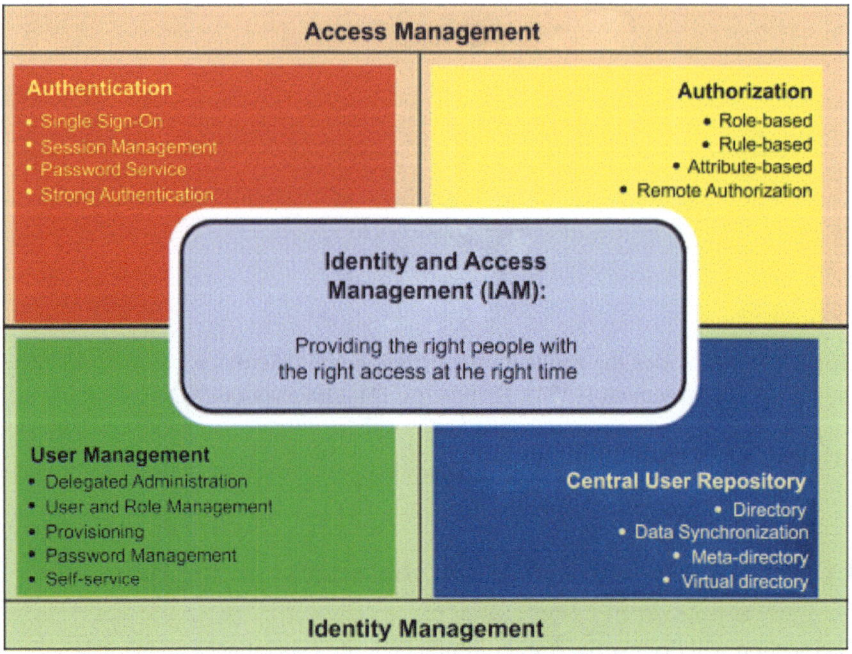

Fig. 4.1 The IAM mitigating approach for cyberattacks on big data

Table 4.1 IAM algorithm

1. Start
2. Define Scope and Objectives
3. Assess Current State
4. Identify User Roles and Privileges
5. Select IAM Solution
6. User Identity Management
7. Role-Based Access Control (RBAC)
8. Data Encryption
9. Access Request and Approval
10. Auditing and Monitoring
11. Regular Review and Maintenance
12. Training and Awareness
13. Incident Response Plan
14. Continuous Improvement
15. End

4.1.1.1 Results of IAM Implementation

- ***Enhanced Data Security:*** IAM implementation establishes stringent access controls and authentication mechanisms, significantly reducing the chances of unauthorized data access. This results in heightened data security and confidentiality.
- ***Reduced Attack Surface:*** The potential attack surface diminishes by enforcing the principle of least privilege through Role-Based Access Control (RBAC). Users only access the data they require, minimizing avenues for cyberattacks.
- ***Improved Compliance***: IAM ensures access and data handling align with regulatory requirements. Compliance with data protection regulations becomes streamlined and demonstrable.
- ***Real-time Monitoring and Alerts:*** Auditing and monitoring functionalities within IAM provide real-time insights into user activities. Suspicious behaviour triggers alerts, enabling rapid responses to potential threats.
- ***Data Integrity:*** Encryption safeguards data in storage and transit, maintaining its integrity even in cyberattacks.

4.1.1.2 Discussion on IAM Implementation

A successful implementation of IAM necessitates the achievement of a delicate equilibrium between security and usability, the facilitation of cultural shifts, the allocation of investments and resources, the consideration of scalability and integration, and the implementation of future-proofing measures. It is important to ensure that employees are adequately informed on the need of IAM compliance, as well as the significance of using secure data handling techniques [6–8]. The topic of centralization vs decentralization is a subject of debate, with advocates expressing support for both perspectives. An essential aspect of this approach is ensuring that IAM solutions are designed to be adaptable to evolving cybersecurity environments and advancements, thus ensuring their long-term viability. Finally, the effective implementation of IAM is contingent upon comprehensive training for end-users. The use of IAM yields many advantages, including enhanced data security, reduced attack surfaces, improved compliance adherence, real-time monitoring capabilities, and fortified data integrity [9, 10]. Nevertheless, the implementation of IAM necessitates the establishment of regular internal communication and the ability to adapt to changing circumstances. The mitigation of cybersecurity concerns pertaining to Big Data might potentially be achieved via the ongoing discourse and adaptability of IAM practices.

4.1.2 Symmetric Data Encryption (SDE)

Symmetric data encryption is a fundamental aspect of contemporary cybersecurity, securing data during transmission or storage. Despite its limits, its rapidity, efficacy, and capacity for ensuring data secrecy make it indispensable for safeguarding sensitive information across many applications. The SDE algorithm is illustrated in Table 4.2.

4.1.2.1 Results of SDE Implementation

SDE has been shown to offer substantial outcomes in enhancing data security and ensuring confidentiality. The following are the primary results of implementing Symmetric Data Encryption.

- ***Enhanced Data Security:*** The primary result of SDE implementation is the enhanced security of sensitive data. Encrypted data is transformed into an unreadable format, ensuring that even if unauthorized parties gain access to the ciphertext, they cannot decipher the original content without the decryption key.
- ***Confidentiality:*** SDE ensures confidentiality by preventing unauthorized individuals from understanding the data. This is particularly crucial when transmitting sensitive information over unsecured networks or storing it in vulnerable environments.
- ***Secure Data Transmission:*** With SDE, data can be securely transmitted over public networks. Even if intercepted during transmission, the encrypted data remains unreadable without the decryption key.
- ***Protection from Cyberattacks:*** Encrypted data is resilient against cyberattacks such as data breaches and eavesdropping. Even if an attacker gains access to the ciphertext, decrypting it without the appropriate key is nearly impossible.

Table 4.2 Symmetric data encryption algorithm

1. Start
2. Generate Encryption Key
3. Prepare Data
4. Divide Data into Blocks
5. Encryption Process
6. Combine Encrypted Blocks
7. Store or Transmit Ciphertext
8. Receive Ciphertext
9. Decrypt Ciphertext
10. Combine Decrypted Blocks
11. Retrieve Original Data
12. End

- ***Regulatory Compliance:*** SDE plays a crucial role in meeting regulatory requirements for data protection and privacy. It ensures that sensitive information is appropriately safeguarded, reducing the risk of compliance violations.
- ***Safe Data Storage:*** Encrypted data remains secure even if physical storage devices are lost or stolen. Unauthorized access to the stored data is thwarted without the decryption key.
- ***Minimal Performance Impact:*** Modern symmetric encryption algorithms are designed for efficiency. The implementation of SDE often has minimal impact on system performance, making it suitable for various applications.

4.1.2.2 Discussion on SDE Implementation

Although SDE offers substantial security advantages, proper key management is essential. To avoid illegal decryption, it is crucial to guarantee the safe development, distribution, and storage of encryption keys. Improved data security, privacy, and resilience in the face of cyberattacks are all visible outcomes of adopting SDE. To reliably share sensitive information while minimizing the dangers of unwanted access and data breaches, SDE has become an integral part of current cybersecurity strategies due to its ability to encrypt data during transmission and storage.

4.1.3 Network Segmentation

The network segmentation is illustrated in Table 4.3.

4.1.3.1 Results of SDE Implementation

Implementing network segmentation has several notable outcomes, greatly improving cybersecurity and network administration. The most important results of implementing network segmentation are as follows:

Table 4.3 Symmetric data encryption algorithm

1. Start
2. Identify Network Assets
3. Categorize Assets by Criticality and Sensitivity
4. Define Segmentation Goals and Policies
5. Design Network Segmentation Architecture
6. Implement Physical or Logical Segmentation
7. Set Up Firewalls and Access Controls
8. Monitor and Analyze Network Traffic
9. Regularly Review and Adjust Policies
10. End

- The major effect of network segmentation is an increase in the network's security. The network's attack surface may be reduced by splitting it into smaller pieces. This restricts attackers' ability to move laterally throughout the network and mitigates their damage.
- Asset Isolation Segmenting a network allows for separating mission-critical resources and data from the rest of the network. This guarantees that the danger of data breaches is reduced even if a single section is compromised.
- The fast spread of threats is slowed by network segmentation in the case of a security breach or hack. Intruders can't spread their damage laterally throughout the network since they can't leave the section they infiltrated.
- Access permissions may be more precisely managed thanks to network segmentation. Each subset may have its access restrictions, limiting users to only those tools they need to do their jobs.
- Organizations that deal with sensitive data or are subject to regulatory compliance might benefit from network segmentation because it offers a systematic way to keep that data separate and secure following best practices. Because network traffic is contained inside its segment, the overall performance of a segmented network may be enhanced. The result is less lag and congestion in the network.
- The attack surface may be minimized by separating individual components. Attackers still encounter challenges when exploiting vulnerabilities inside a segment, even if they exist.

4.1.3.2 Discussion on SDE Implementation

Recognizing that the process of network segmentation calls for careful planning and ongoing management is an essential step in the endeavor. For the effectiveness of the segmented network to be maintained over time, it is important to perform diligent monitoring, update and modify policies on a consistent basis, and make any necessary policy adjustments [11, 12]. The decision of an organization to implement network segmentation has a significant bearing on the level of overall cybersecurity preparedness that the organization has. Organizations should prepare for a wide variety of significant repercussions, such as increased security measures, improved asset isolation, more effective threat containment, and improved access control. The ability to properly handle the potential risks connected with cyberattacks is made possible for businesses by segmenting their networks. This secures sensitive data and maintains the integrity of the design of a network [13–15].

4.2 Proposes Integrated Approach

This section discusses and proposes an integrated approach to mitigate the risk of cyberattacks on big data. It is vital to integrate several tactics to successfully minimize the risks of cyberattacks on Big Data in the current cybersecurity environment, which

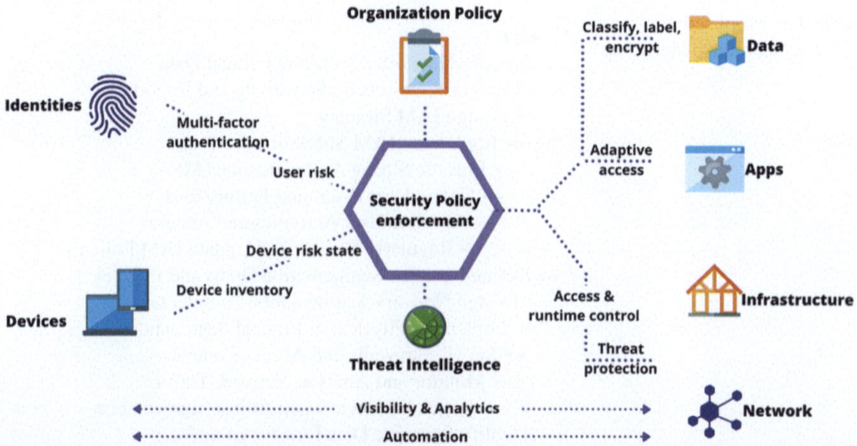

Fig. 4.2 Proposed integrated approach architecture

is constantly growing. The following is an all-encompassing strategy that combines several preventative measures to protect Big Data from possible dangers.

The proposed integrated approach implements "IAM" to control user access and privileges by enforcing strong authentication methods, such as multi-factor authentication, to ensure users have the least right necessary for their roles. It also uses "Network Segmentation" to divide the network into segments based on asset criticality and sensitivity by isolating sensitive Big Data repositories from less critical parts of the network and applying strict access controls between segments to limit lateral movement. Finally, it applies SDE to encrypt data at rest and during transmission by using strong encryption algorithms and key management practices and encrypting sensitive Big Data to ensure confidentiality and integrity.

Thus, Fig. 4.2 illustrates the architecture and Table 4.4 shows the proposed integrated approach algorithm.

4.2.1 Result and Discussion of the Proposed Integrated Approach

IAM is essential for ensuring that only authorized individuals may access critical data. It does this by implementing stringent authentication techniques, role-based access restrictions, and the concept of least privilege, which ensures that only authorized users may access sensitive data. Network segmentation involves dividing the network into smaller sections, isolating important assets and restricting lateral movement. This prevents illegal lateral movement across separate network segments, which limits the amount of harm that might be done [12, 16]. By ensuring that encrypted data remains unreadable even if illegal access is gained, SDE adds another line of protection against

Table 4.4 Proposed integrated approach algorithm

1. Start
→ Identify Network Assets and Critical Data
→ Categorize Assets by Sensitivity and Importance
→ Design IAM Strategy
→ Implement IAM Solution
→ Enforce Strong Authentication (MFA)
→ Define User Roles and Permissions
→ Monitor User Activities and Access
→ Regularly Review and Update IAM Policies
→ Define Network Segmentation Goals and Policies
→ Design Network Segmentation Architecture
→ Implement Physical or Logical Segmentation
→ Set Up Firewalls and Access Controls
→ Monitor and Analyze Network Traffic
→ Regularly Review and Adjust segmentation
→ Identify Sensitive Data for Encryption
→ Choose Symmetric Encryption Algorithm
→ Generate Encryption Keys
→ Apply Data Encryption
→ Securely Store Encryption Keys
→ Establish Key Management Practices
2. End

the possibility of data breaches. Organizations can create a multi-layered defensive strategy by combining various tactics, with each layer enhancing the qualities of the others. IAM determines who can access data, Network Segmentation restricts mobility within the network, and SDE ensures that it does not become readable even if data is accessible. The organization's total security posture is greatly improved due to this synergy, reducing the chance of successful assaults and data breaches [17, 18].

Granular access management is a mechanism that uses identity access management (IAM) and network segmentation to govern who has access to what resources and information. This may be accomplished by segmenting the network. IAM reduces the risk that unauthorized parties will obtain confidential information by ensuring that users have access to just the data necessary to carry out their jobs. With the help of segmentation, you can prevent users from moving laterally across the network and limit their access to the resources they need. This all-encompassing strategy makes it possible to regulate data access with more granularity, enhance data privacy, lower risk, make regulatory compliance more straightforward, and raise user productivity [19]. Granular access control may increase security, but only if it is done with foresight, access laws are reviewed often, and users are kept in the loop. Otherwise, the deposit may not improve. The roles and responsibilities of users must be accurately represented in the access restrictions. Therefore, a better data environment may be constructed using IAM and Network Segmentation to give fine-grained access control. This can be accomplished. This reduces the surface area that might be exploited and complies with all applicable privacy regulations at the same time [20].

The incorporation of IAM, Network Segmentation, and SDE into an organization's security architecture might help it become more resistant to attacks from inside. IAM ensures that only authorized persons can access private information, preventing unwanted outsiders or dishonest insiders from getting unauthorized access. IAM assures that only authorized individuals may access confidential information in this approach [21]. By separating the different parts of the network, network segmentation prevents attackers from inside the network from taking advantage of its flaws and propagating laterally across the web. By making it hard to read encrypted data in the absence of the appropriate key, symmetric data encryption provides an additional layer of defence against unauthorized individuals who may be present inside the network. Because it consists of many layers, this method has significantly reduced the amount of damage that a breach on the inside may cause. A few benefits of using this strategy include ensuring the safety of one's data, limiting exposure, establishing trust, and complying with regulations [22]. Managing internal hazards requires striking a balance in the workplace between protecting employees and maximizing their output. To prevent employees from feeling alienated, a company's reaction to an occurrence and the following adoption of safeguards need to be handled with care. To summarize, integrating Identity and Access Management (IAM), Network Segmentation, and Secure Data Exfiltration (SDE) protects against unauthorized internal access and lateral movement inside the network. This mitigates the risk of unintentional data breaches and hostile insiders and helps create an atmosphere where data can be trusted [23].

IAM, Network Segmentation, and SDE working together provide a powerful defensive mechanism that ensures the integrity of the other levels even if a vulnerability is discovered in one of the layers. It is of the utmost importance to strike a balance between security and usability requirements. If this harmony can be achieved, solid security measures may be implemented while at the same time user procedures are disrupted as little as possible [24]. Nevertheless, in order to properly integrate these components, careful strategic planning and execution are required, which must include the elimination of any technological, operational, and instructional barriers. The purchase, implementation, and training that are linked with technology may give rise to concerns relating to cost. Because the integrated approach exhibits a solid agreement with regulatory regulations regarding data security and privacy, it makes the process of demonstrating compliance much easier. The capability to scale is an essential component of the integrated strategy because it enables alignment with the progression of the business and the ever-shifting threat environment. In order to maintain this integrated methodology's usefulness over time, regular reviews and adjustments are necessary [24, 25]. In conclusion, merging IAM, Network Segmentation, and SDE into a cybersecurity strategy offers a holistic approach that strengthens data protection, regulates access rights, and fortifies resistance against harmful cyber activities. These are all important components of a comprehensive cybersecurity strategy.

4.2.2 Critical Analysis

Cyberattacks on large amounts of data might have devastating effects, including threats to national security, monetary losses, breaches of users' privacy, disruptions to business operations, and theft of intellectual property. To mitigate the effects of these conceivable dangers, companies have the choice to implement any number of preventative measures. Encryption of data, access control and authentication mechanisms, continuous monitoring practices, regular patching and updates, employee training and awareness initiatives, backup and recovery plans, cyber insurance coverage, and approaches to collaboration and information sharing are all included in these strategies. Data breaches may be caused by cyberattacks that target big data sets [26]. This can result in the disclosure of personal information and compromise an individual's right to privacy, both of which can have legal repercussions. There is a possibility that the company may face financial losses as a result of the expenditures required for the cleaning, the legal fees, and the negative influence on the company's image. An operational interruption may affect several parts of an organization, including supply chains, customer service, and the continuation of the firm as a whole [27]. Intellectual property theft may cause a decline in a company's advantage over its competitors and should be avoided at all costs. Cyberattacks directed against huge datasets can potentially compromise the security of sensitive information, intelligence operations, and crucial infrastructure within government and national security agencies [28].

Companies can use data encryption to safeguard data throughout the transmission and storage stages of the data life cycle to mitigate the risks that these possible threats may pose. Access control techniques and multi-factor authentication may be necessary to limit the danger of unauthorized access to large-scale data repositories. Continuous monitoring systems enable the power to quickly detect and respond to potential cyber threats. As a preventative precaution against the exploitation of vulnerabilities by hackers, the deployment of routine software and system patching and updates is an important preventive measure. Introducing staff training programs and developing awareness among workers may make recognizing cyber hazards easier and responding appropriately [29]. It's possible that implementing data backups regularly and developing thorough recovery plans may successfully decrease the danger of data loss and cut down the amount of time an entire system is down. If a cyberattack is successful, deploying cyber insurance policies can mitigate the financial losses that occur due to the assault. Collaboration and information sharing across colleagues in the sector may make it easier to take preventative measures against attacks [30].

In general, the effects of cyberattacks on huge amounts of data may have significant repercussions. Nevertheless, businesses can effectively mitigate the risks associated with such assaults by implementing robust cybersecurity measures and being continually alert. To keep up with the always-shifting landscape of potential dangers, those working in cybersecurity must maintain a firm commitment to adaptation.

Protecting big data is necessary for maintaining the confidentiality of sensitive information, guaranteeing the smooth running of operations, and safeguarding national security for individuals, corporations, and governments.

4.3 Mitigation Strategies

Mitigation strategies include proactive actions and approaches businesses may use to diminish the likelihood of security events and mitigate the possible consequences of security breaches. Implementing these techniques is paramount in protecting data, systems, and networks [31]. The following are a few essential techniques for mitigation.

1. **Access Management:** It involves;
    - User Authentication: Using robust authentication techniques, such as multi-factor authentication (MFA) and biometrics, is crucial to establish that only authorized individuals are granted access to systems and data [32].
    - Role-Based Access Control (RBAC) is a security mechanism that assigns distinct roles and permissions to individual users, restricting their access to just the resources and functionalities essential for performing their job responsibilities [33].
    - The Least Privilege Principle is a security measure that provides people with the lowest possible access necessary to carry out their assigned activities. This practice minimizes the potential for unauthorized access and associated risks [34].
2. **Cryptography:** Data encryption is a crucial security measure involving the transformation of sensitive data, both while it is stored and when it is being sent, to safeguard it from unwanted access. Using robust encryption methods and adhering to effective key management policies is essential. End-to-end encryption ensures data security throughout its lifecycle, from when it is generated to when the intended recipient accesses it [35].
3. **Network Security:** Firewalls regulate the flow of network traffic, both incoming and outgoing, to prevent unwanted access and possibly harmful data from entering or leaving a network. Intrusion Detection and Prevention Systems (IDS/IPS) are used to actively monitor network traffic to detect any potentially malicious behavior and initiate automated responses to counteract identified threats [36, 37].
4. **Frequent Updates and Patching:** It is essential to ensure that all software, operating systems, and applications are regularly updated with security updates to mitigate known vulnerabilities effectively. Attackers often exploit vulnerabilities [38].

5. **Security Awareness and Education:** The objective is to provide workers with instruction on security best practices, enhance their understanding of social engineering tactics, and enable them to identify and thwart phishing efforts to mitigate the risk of unauthorized access.
6. **Plan of Action for an Incident:** Establishing and maintaining an incident response strategy is essential to guarantee a prompt and efficient reaction to security occurrences. It is important to conduct periodic testing of the strategy to verify its efficacy in practical use [38].
7. **Vulnerability Control:** It is important to conduct routine scans of systems and applications to identify potential vulnerabilities. These vulnerabilities should then be assessed and prioritized according to their level of risk. Subsequently, urgent action must be taken to address and resolve these vulnerabilities via remediation or mitigation measures.
8. **Network Segmentation**: One effective strategy to mitigate the potential impact of attackers who have successfully infiltrated a certain network section is to implement network segmentation. By dividing the network into distinct segments, the lateral movement of attackers may be restricted, hence minimizing the extent of their potential damage or unauthorized access. This measure may effectively mitigate security breaches [39].
9. **Security Surveillance:** Utilize security information and event management (SIEM) systems with log analysis techniques to monitor and detect potentially malicious activity and aberrations in real time [40].
10. **Restore and Back Up:** It is important to consistently create backups of essential data and systems and thoroughly test the restoration process. This practice is crucial to maintain business continuity during data breaches or system failures.

The implementation of an effective security system necessitates the use of a variety of mitigation measures, which should be customized to align with the unique requirements and risk profile of a company. Regular security evaluations and testing are necessary to maintain the ongoing effectiveness of these safeguards. The maintenance of security requires a continuous cycle of surveillance and enhancement.

4.3.1 Data Encryption and Access Control

Encryption of data and authorization and authentication of users are essential parts of information security. They are essential in preventing unauthorized access to data and preserving the security and integrity of sensitive information, which are very important [41]. Both of these ideas are examined in further detail below.

1. **Encryption of Data:** Encryption is turning data from its plaintext form into its ciphertext form to prevent unauthorized access to the data. It requires using cryptographic techniques and keys to render data unreadable to anybody who does not possess the appropriate decryption key [26]. The following outlines some important aspects of data encryption: Encryption guarantees that even if

an unauthorized person can access the encrypted data, they cannot understand it without the encryption key. This protects the confidentiality of the data. Different kinds of encryption exist, such as symmetric and asymmetric encryption, for example, and these are only two of the many options. In symmetric encryption, the same key is used for both the encryption and the decryption processes, but in asymmetric encryption, a pair of public and private keys is used instead [27].

- **Use Cases:** Encryption. may be used to safeguard data while it is at rest (stored on discs or in databases), data while it is in transit (while it is being sent across networks), and data while it is being utilized (while it is being processed or used by applications).
- **End-to-end Encryption:** End-to-end encryption is a method that assures data continues to be encrypted from the point of creation to the point of consumption, even while the data is being transferred via the services or systems of a third party.
- **Management of the Keys:** Ensuring that the appropriate management of the keys is necessary to preserve the confidentiality of encrypted data. It involves creating robust keys, keeping them in a safe location, and rotating them at appropriate intervals.

Compliance with legislation: There are a variety of data protection legislation and standards, such as GDPR and HIPAA, that mandate encrypting sensitive data to satisfy compliance requirements.

2. Access Management

Access control manages and regulates who can access certain resources, such as data, systems, and physical places. It include managing and controlling who has access to specific resources. It includes authentication and authorization as parts of its scope [42]. The following are some important aspects that pertain to access control.

Authentication is the process of confirming the identity of a person or system. Authentication may refer to any of these. Typically, this is accomplished via several authentication methods, including passwords, biometrics, smart cards, and multi-factor authentication (MFA) [31]. Authorization is the process that comes after authentication and includes allowing or refusing access to certain resources depending on the user's identification and the rights they have been granted. Authorization comes after authentication. These policies govern access control, and they are enforced here. RBAC, or role-based access control, is a standard method of access control that assigns users to roles and provides rights depending on those roles. RBAC is an abbreviation for role-based access control. Management of access controls is made easier as a result.

The Least Privilege Principle states that users and systems should only be provided the least access required to execute the assigned tasks to minimize the possibility of unauthorized access [43, 44].

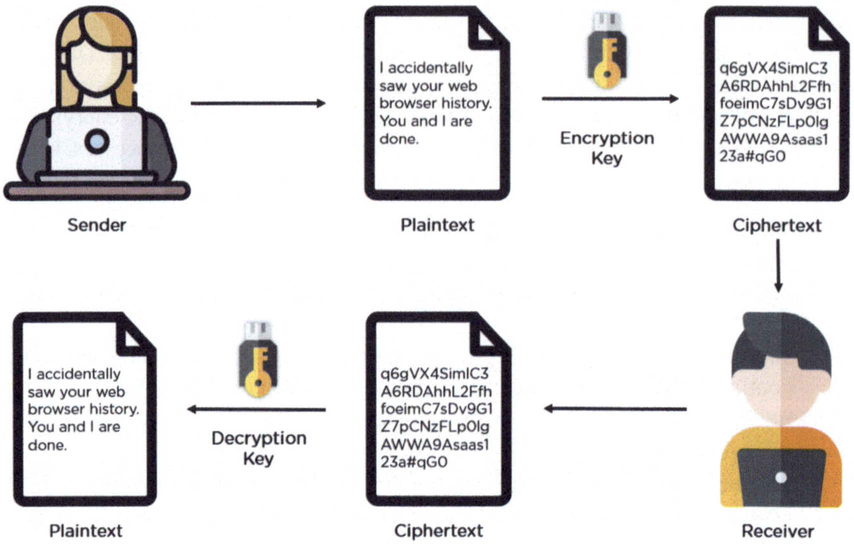

Fig. 4.3 Data Encryption and Access Control

- **Access Management Lists (ACLs):** Access Control Lists (ACLs) indicate which people or systems can access specified resources. It enables more fine-grained management of the resources.
- **Monitoring of Access:** Access control systems should include monitoring and tracking access attempts to identify suspicious or unauthorized behavior.
- **Controlling Physical Access**: Besides controlling logical access, businesses and other organisations need to regulate physical access to data centres and other sensitive places to prevent unauthorized physical breaches.

Encryption of data and authorization of users are two crucial elements that must be included in any complete security plan. Data encryption prevents it from being seen by unauthorized parties, even if a security breach occurs. At the same time, access control guarantees that only authorized people or systems may interact with data and resources. These precautions, in conjunction with one another, will reduce the likelihood of data breaches and safeguard confidential information, as illustrated in Fig. 4.3.

4.3.2 Regular Data Backups and Disaster Recovery Plans

Regular data backups and disaster recovery plans are critical to an organization's overall data management and security strategy. They help ensure the availability and integrity of data, even in the face of unexpected events or data loss. An overview of both concepts includes:

1. **Regular Data Backups**

Regular data backups involve creating duplicate copies of critical data and storing them in a secure location. The primary purpose of backups is to provide a means to recover data in case of data loss, corruption, or other unexpected incidents. Here are key points related to regular data backups.

- **Data Types**: Backups can encompass various data types, including databases, files, configurations, and system images.
- **Frequency**: Data backups should be performed regularly, with the frequency depending on the organization's needs and the criticality of the data. For some systems, this could be daily, weekly, or even in real-time.
- **Data Retention**: Organizations need to determine how long backups should be retained. This may be based on legal requirements, compliance regulations, or business needs.
- **Methods**: Backup methods include full backups (copying all data), incremental backups (copying changes since the last backup), and differential backups (copying changes since the last full backup).
- **Offsite Storage**: Backups should be stored in a secure, offsite location to protect against on-site disasters, such as fires or floods.
- **Testing**: Regularly test the restoration process to ensure that backups are viable and data can be recovered successfully.

2. **Disaster Recovery Plans**

A disaster recovery plan (DRP) is a comprehensive strategy that outlines how an organization will respond to and recover from significant disruptions to its operations. This plan includes data recovery, processes, and procedures to ensure business continuity [45]. The key points related to disaster recovery plans are;

- **Risk Assessment**: Identify potential risks and threats that could lead to data loss or system downtime. It includes natural disasters, cyberattacks, hardware failures, and more.
- **Prioritization**: Determine the critical systems and data that must be recovered first. Establish recovery time objectives (RTOs) and recovery point objectives (RPOs) to define how quickly data and systems should be restored.
- **Backup Strategy**: The DRP should specify the backup and recovery strategies, including data backup frequency, retention, and restoration procedures.
- **Communication Plan**: Define how communication will be managed during a disaster, both internally and externally, to keep stakeholders informed.
- **Testing and Training**: Regularly test the DRP and train employees on their roles and responsibilities during a disaster. It ensures that everyone knows what to do in case of an emergency.
- **Alternative Facilities**: Identify alternative facilities or cloud-based resources that can be used to maintain operations if the primary location is unavailable.
- **Documentation**: Document the entire disaster recovery plan, including contact information, procedures, and any necessary resources.

- **Continuous Improvement**: Regularly review and update the DRP to adapt to changing risks, technologies, and business needs.

Disaster recovery plans are essential for minimizing downtime and data loss in unexpected events, whether natural disasters, cyberattacks, or other disruptions. When combined with regular data backups, organizations can significantly reduce the impact of such incidents on their operations and data integrity.

4.3.3 Intrusion Detection and Prevention Systems

Network security relies heavily on IDS and Intrusion Prevention Systems (IPS). While IDS is meant to identify and notify of suspicious activity in a network, intrusion prevention systems (IPS) not only detect but also actively prevent or stop such actions from occurring in the network. IDS uses a variety of detection approaches, some of which include signature-based detection, anomaly-based detection, and heuristic or behavioral analysis. The difference between signature-based and anomaly-based detection is that the former examines current behavior concerning a baseline, while the latter matches predetermined patterns or signatures of known assaults. The heuristic approach looks for patterns of behavior to detect potentially risky behaviors [46]. IDS are responsible for generating alarms, which security professionals then examine and analyze. IDS may be implemented as either a Network-Based IDS (NIDS), which monitors network traffic, or a Host-Based IDS (HIDS), which monitors actions on individual computers. Both types of IDS are referred to as IDS. In most cases, an IDS will function passively, which entails monitoring and alerting rather than actively attempting to prevent intrusions [47].

IPS makes use of the same detection mechanisms as IDS, but in addition to that, it takes preventative or blocking actions. They can quarantine questionable traffic, reset connections, block malicious traffic, notify of dangers, and actively guard certain systems. Network-Based IPS (NIPS) and Host-Based IPS (HIPS) are two different ways IPS may be implemented. In contrast to IDS, IPS are proactive security solutions that can carry out automatic operations in real-time. Because they offer early warning and reaction mechanisms against cyber attacks, intrusion prevention systems (IPS) and intrusion detection systems (IDS) are essential for network security. The level of risk an organization is willing to take, and its security needs should determine whether it deploys an IPS, IDS, or a mix of the two [48]. IDS, in conclusion, places its primary emphasis on monitoring and alerting, while IPS goes one step further by proactively avoiding or blocking potential threats. Together, they are a crucial component of the architecture that makes up the network's security, defending against actions that have not been authorized or are harmful, as shown in Fig. 4.4.

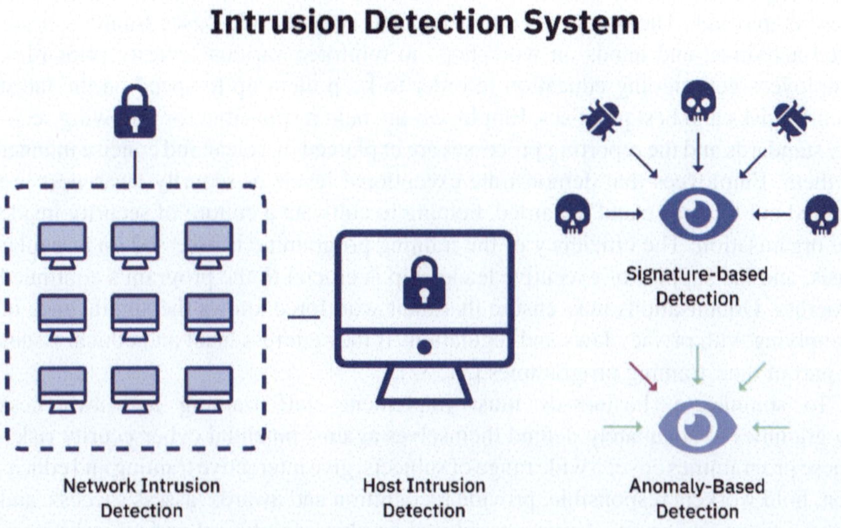

Fig. 4.4 Intrusion detection and prevention systems

4.3.4 *Employee Training and Awareness Programs*

Training and awareness programs for employees are very necessary for businesses to implement in order to defend themselves from potential cybersecurity risks. These programmes educate participants on a variety of subjects, including phishing awareness, managing passwords, social engineering, malware awareness, safe online surfing, data handling, mobile device security, and physical security. The participatory training involves role-based training, simulated phishing activities, and hands-on workshops [49]. Education on an ongoing basis is offered to staff members to keep them current on the latest security best practises and emerging dangers. Employees are held responsible for following security standards, and the reporting processes are explained clearly and concisely. Employees who demonstrate exceptional security awareness are lauded and rewarded for their efforts. The efficiency of the training programme is measured consistently, and the executive team's support is essential. The company instills in its employees a sense that protecting sensitive data is everyone's duty and actively works to foster this culture [50]. In addition, discussions on legal and ethical issues will occur throughout the course.

In general, employee training and awareness programmes play a critical role in ensuring that workers are educated and empowered to recognise and react appropriately to cybersecurity risks. This may be accomplished by ensuring that employees are able to recognise and respond appropriately. The workers of an organisation may be equipped with the knowledge and skills necessary to secure the organization's data and systems if the organisation provides training on a variety of subjects to those employees. The training is interactive and involves role-based training, simulated activities, and hands-on workshops to reinforce various security principles. Employees get ongoing education in order to keep them up to speed on the latest security risks and best practises. Employees are held responsible for following security standards and the reporting processes are explained in a clear and concise manner to them. Employees that demonstrate exceptional levels of security awareness are singled out for praise and rewarded, helping to cultivate a culture of security inside the organisation. The efficiency of the training programme is assessed on a regular basis, and the support of executive leadership is crucial to the program's continued progress. Organisations may ensure that their workforce knows the significance of complying with privacy laws and regulations if they address legal and ethical issues as part of their training programmes [51, 52].

To summarise, businesses must implement staff training and awareness programmes to adequately defend themselves against potential cybersecurity risks. These programmes cover a wide range of subjects, give interactive training and education, hold workers responsible, provide recognition and awards, assess success, and place a focus on executive support, cultural emphasis, and legal and ethical issues. Through implementing these programmes, organisations can guarantee that their staff are well-informed and alert, which contributes to a better safe environment inside the organization.

4.3.5 Compliance with Cybersecurity Regulations

Compliance with cybersecurity regulations is crucial for organizations across industries to protect sensitive data, maintain information system integrity, and ensure individual privacy. Non-compliance can lead to severe consequences, including fines, legal action, and reputational damage [53]. To achieve compliance, organizations should follow these key aspects.

1. **Understand applicable regulations:** Identify the cybersecurity regulations that apply based on industry, location, and data handled, such as GDPR, HIPAA, PCI DSS, and national data protection laws.
2. **Conduct compliance assessment:** Assess current cybersecurity practices and infrastructure to identify gaps and areas of non-compliance.
3. **Classify data:** Categorize data to determine which is subject to regulatory requirements, including sensitive personal information, financial data, and intellectual property.
4. **Protect data**: Implement data protection measures like encryption, access controls, and data loss prevention to safeguard sensitive data as regulations require.
5. **Establish security policies and procedures**: Develop, document, and enforce cybersecurity policies, procedures, and standards-aligned with regulatory requirements.

6. **Create a data breach response plan:** Establish an incident response plan to address data breaches and comply with reporting requirements promptly.
7. **Implement access control:** Use role-based access control to ensure users have appropriate access permissions based on their roles and responsibilities.
8. **Provide security awareness training:** Regularly train employees on cybersecurity regulations, organizational policies, and their compliance roles.
9. **Conduct regular audits and assessments:** Perform internal and external audits and assessments to evaluate compliance with cybersecurity regulations.
10. **Assess vendors and third parties:** Evaluate the security practices of third-party vendors and service providers with access to data to ensure compliance.
11. **Understand reporting requirements**: Comply with reporting and notification requirements in case of data breaches or security incidents.
12. **Implement privacy by design:** Consider data protection and privacy throughout the development and lifecycle of systems and applications.
13. **Manage data retention and destruction**: Follow regulations on data retention and secure destruction, ensuring data is not retained longer than necessary.
14. **Ensure data subject rights:** Comply with data subject rights, such as access, correction, or deletion of personal information, as regulations specify.
15. **Stay updated on regulatory changes**: Stay informed about changes to cybersecurity regulations that may impact compliance requirements.
16. **Maintain documentation and records:** Keep detailed records and documentation of cybersecurity efforts, assessments, and compliance activities.
17. **Seek legal expertise:** Consider legal counsel or cybersecurity experts specializing in regulatory compliance to ensure meeting all legal requirements.

Compliance with cybersecurity regulations requires ongoing dedication, resources, and a commitment to data protection. Organizations that prioritize compliance not only avoid legal consequences but also demonstrate their commitment to protecting customers and stakeholders.

4.4 Chapter Summary

This chapter explains how to strengthen cybersecurity against current threats by integrating Identity and Access Management (IAM), Network Segmentation, and Symmetric Data Encryption (SDE). When these tactics are combined, businesses can strengthen their data security, implement granular access control, and become more resistant to internal and external attacks. While IAM regulates user access and privileges, network segmentation ensures users cannot move laterally across the network. The use of symmetric data encryption ensures that private information is rendered unintelligible if it is accessed in an unauthorized manner. This chapter strongly emphasizes the power of integration, which creates a multi-layered defense mechanism that magnifies the advantages of each method. Complexity, continuing management, and the user experience are some of the challenges mentioned here.

Integration of Identity and Access Management (IAM), Network Segmentation, and Security Data and Event Management (SDE) enables businesses to obtain fine-tuned control over data access, limit possible attack surfaces, and protect themselves from internal and external threats.

References

1. Hunter, W.C.: J. Smart Tour. **1**(2), 27–36 (2021). http://smarttourism.khu.ac.kr/file/202103/1622686933.pdf
2. Song, S.K., et al.: Prescriptive analytics system for improving research power. In: Proceedings of 16th International Conference on Computational Science and Engineering (CSE 2013), pp. 1144–1145 (2013). https://doi.org/10.1109/CSE.2013.169.
3. Álvarez-Monzoncillo, J.M.: The Dynamics of Influencer Marketing: A Multidisciplinary Approach (2022). https://doi.org/10.4324/9781003134176
4. Sánchez, M.: A general approach on privacy and its implications in the digital economy. J. Econ. Issues **56**(1), 244–258 (2022). https://doi.org/10.1080/00213624.2022.2025729
5. Scianna, A., Gaglio, G.F., La Guardia, M.: Structure monitoring with BIM and IoT: the case study of a bridge beam model. ISPRS Int. J. Geo-Inf. **11**(3), 102–113 (2022). https://doi.org/10.3390/ijgi11030173
6. Alkahtani, H., Aldhyani, T.H.H.: Artificial intelligence algorithms for malware detection in android-operated mobile devices. Sensors **22**(6), 1–26 (2022). https://doi.org/10.3390/s22062268
7. Kansagra, D., Kumhar, M., Jha, D.: Ransomware: a threat to cyber security. Comput. Sci. Electron. J. **7**(1), 224–227 (2016). https://doi.org/10.090592/IJCSC.2016.035
8. Kumar N.M., et al.: Distributed energy resources and the application of AI, IoT, and blockchain in smart grids. Energies **13**(21) (2020). https://doi.org/10.3390/en13215739
9. Wang, J., Paschalidis, I.C.: Botnet detection based on anomaly and community detection. IEEE Trans. Control Netw. Syst. **4**(2), 392–404 (2017). https://doi.org/10.1109/TCNS.2016.2532804
10. Ambwani, T.: Essential cybersecurity controls (ECC-1:2018) standard compliance. Saudi Arab. Natl. Cybersec. Auth. 1–15 (2018)
11. Ohri, K., Kumar, M.: Review on self-supervised image recognition using deep neural networks. Knowl. Based Syst. **224**, 107090 (2021). https://doi.org/10.1016/j.knosys.2021.107090
12. Pambhar, H., Aghera, K., Tada, N., Residual, Á.Á.: An Advanced Web-Based Bilingual Domain Independent Interface to Database Using Machine Learning Approach, vol. 508, no. June, pp. 197–204 (2018). https://doi.org/10.1007/978-981-10-2750-5
13. Sengupta, J., Ruj, S., Das Bit, S.: A comprehensive survey on attacks, security issues and blockchain solutions for IoT and IIoT. J. Netw. Comput. Appl. **149**(12), 102–115 (2020). https://doi.org/10.1016/j.jnca.2019.102481
14. Dong, T., Li, S., Qiu, H., Lu, J.: An Interpretable Federated Learning-based Network Intrusion Detection Framework, pp. 1–12 (2022). http://arxiv.org/abs/2201.03134
15. Tufail, S., Parvez, I., Batool, S., Sarwat, A.: A survey on cybersecurity challenges, detection, and mitigation techniques for the smart grid. Energies **14**(18), 1–22 (2021). https://doi.org/10.3390/en14185894
16. Hassan, B., et al.: Deep learning based joint segmentation and characterization of multi-class retinal fluid lesions on OCT scans for clinical use in anti-VEGF therapy. Comput. Biol. Med. **136**(July), 104727 (2021). https://doi.org/10.1016/j.compbiomed.2021.104727
17. Saad, M., Bin Ahmad, M., Asif, M., Khan, M.K., Mahmood, T., Mahmood, M.T.: Blockchain-enabled VANET for smart solid waste management. IEEE Access **11**(November), 5679–5700 (2023). https://doi.org/10.1109/ACCESS.2023.3235017

18. Siddiqui, S., Hameed, S., Shah, S.A., Khan, A.K., Aneiba, A.: Smart contract-based security architecture for collaborative services in municipal smart cities [formula presented]. J. Syst. Archit. **135**(December), 102802 (2023). https://doi.org/10.1016/j.sysarc.2022.102802
19. Catalano, C., Paiano, L., Calabrese, F., Cataldo, M., Mancarella, L., Tommasi, F.:Anomaly detection in smart agriculture systems. Comput. Ind. **143**(December), 103750 (2022). https://doi.org/10.1016/j.compind.2022.103750
20. Eze, K.G., Akujuobi, C.M., Hunter, S., Alam, S., Musa, S., Foreman, J.: A Blockchain-based security architecture for the internet of things. WSEAS Trans. Inf. Sci. Appl. **19**(5), 12–22 (2022). https://doi.org/10.37394/23209.2022.19.2
21. Cao, J.: Coordinated development mechanism and path of agricultural logistics ecosystem based on big data analysis and IoT assistance. Acta Agric. Scand. Sect. B Soil Plant Sci. **72**(1), 214–224 (2022). https://doi.org/10.1080/09064710.2021.2008476
22. Kemmoe, V.Y., Kwon, Y., Hussain, R., Cho, S., Son, J.: Leveraging smart contracts for secure and asynchronous group key exchange without trusted third party. IEEE Trans. Depend. Secur. Comput. 1–18 (2022). https://doi.org/10.1109/TDSC.2022.3189977
23. Stroumpoulis, A., Kopanaki, E.: Theoretical perspectives on sustainable supply chain management and digital transformation: a literature review and a conceptual framework. Sustainability **14**(8) (2022). https://doi.org/10.3390/su14084862
24. Al-otaibi, S.Z.: Data security challenges with its defence strategies of internet of things: critical review study. Commun. Math. Appl. **13**(1), 401–415 (2022). https://doi.org/10.26713/cma.v13i1.1980
25. Beggar, I., Riahla, M.A.: The internet of things security challenges: survey. Netw. Syst. **4**(13), 356–366 (2022). https://doi.org/10.1007/978-3-030-96311-8_33
26. Hameed, S., et al.: A scalable key and trust management solution for IoT sensors using SDN and blockchain technology. IEEE Sens. J. **21**(6), 8716–8733 (2021). https://doi.org/10.1109/JSEN.2021.3052009
27. Tsiknas, K., Taketzis, D., Demertzis, K., Skianis, C.: Cyber threats to industrial IoT : a survey on attacks and countermeasures. Preprints, no. February, pp. 1–26 (2021). https://doi.org/10.20944/preprints202102.0148.v1
28. Na, D., Park, S.: Fusion chain: a decentralized lightweight blockchain for iot security and privacy. Electron **10**(4), 1–18 (2021). https://doi.org/10.3390/electronics10040391
29. Saleem, N., Rahman, A., Rizwan, M., Naseem, S., Ahmad, F.: Enhancing security of android operating system based phones using quantum key distribution. EAI Endorsed Trans. Scalable Inf. Syst. **7**(28), 1–8 (2020). https://doi.org/10.4108/eai.13-7-2018.165281
30. Cinar, A.C., Kara, T.B.: The current state and future of mobile security in the light of the recent mobile security threat reports. Multimed. Tools Appl. **82**(13), 20269–20281 (2023). https://doi.org/10.1007/s11042-023-14400-6
31. Batista, J.P.B., Torre, C., Sousa Lobo, J.M., Sepodes, B.: A review of the continuous professional development system for pharmacists. Hum. Resour. Health **20**(1), 1–12 (2022). https://doi.org/10.1186/s12960-021-00700-1
32. Sultana, T., Almogren, A., Akbar, M., Zuair, M., Ullah, I., Javaid, N.: Data sharing system integrating access control mechanism using blockchain-based smart contracts for IoT devices. Appl. Sci. **10**(2) (2020). https://doi.org/10.3390/app10020488
33. Karthick, S., Binu, S.: Android security issues and solutions. In: IEEE 2017 International Conference on Innovative Mechanisms for Industry Applications (ICIMIA 2017)—Proceedings, no. August, pp. 686–689 (2017). https://doi.org/10.1109/ICIMIA.2017.7975551
34. Neuman, B.C., Ts'o, T.: Kerberos: an authentication sevice for computer networks. IEEE Commun. Mag. **32**(September), 33–38 (2014). https://doi.org/10.1109/35.312841
35. Zukarnain, Z.A., Muneer, A., Ab Aziz, M.K.: Authentication securing methods for mobile identity: issues, solutions and challenges. Symmetry **14**(4) (2022). https://doi.org/10.3390/sym14040821
36. Almogren, A., Mohiuddin, I., Din, I.U., Almajed, H., Guizani, N.: FTM-IoMT: fuzzy-based trust management for preventing sybil attacks in internet of medical things. IEEE Internet Things J. **8**(6), 4485–4497 (2021). https://doi.org/10.1109/JIOT.2020.3027440

37. Rueda, D.F., Caviedes, J.C., Muñoz, W.Y.C.: A hybrid intrusion detection approach based on deep learning techniques. Lect. Notes Data Eng. Commun. Technol. **117**(March), 863–878 (2022). https://doi.org/10.1007/978-981-19-0898-9_65
38. Meskell, L.: Rainey and the Russians: arctic archaeology, 'eskimology' and cold war cultural diplomacy. Archaeol. Dial. 1–17 (2022). https://doi.org/10.1017/s1380203822000228
39. Min, D., et al.: Amoeba: an autonomous backup and recovery SSD for ransomware attack defense. IEEE Comput. Archit. Lett. **17**(2), 243–246 (2018). https://doi.org/10.1109/LCA.2018.2883431
40. Modgil, S., Gupta, S., Stekelorum, R., Laguir, I.: AI technologies and their impact on supply chain resilience during-19. Int. J. Phys. Distrib. Logist. Manag. **52**(2), 130–149 (2022). https://doi.org/10.1108/IJPDLM-12-2020-0434
41. Hammad, M., et al.: Security framework for network-based manufacturing systems with personalized customization: an industry 4.0 approach. Sensors 23(17) (2023). https://doi.org/10.3390/s23177555
42. Aswir, Misbah, H.: Smartphones as personal digital archives? Recentering migrant authority as curating and storytelling subjects. Photosynthetica **2**(1), 1–13 (2018). https://doi.org/10.1007/978-3-319-76887-8
43. Ntlotlang, T.: Technology Mediated Tools As Drivers of Library-Researcher Collaboration: The Case of Botswana International University of Science and Technology (BIUST) Institutional Repository (IR), pp. 1–10 (2019)
44. Zhi, Z., Abba, N.B., Hamid, A.A.: Employee participation in organizational decision making as a motivational factor for building high performance work system in an organization. Int. J. Innov. Res. Adv. Stud. **7**(5), 111–116 (2020). https://file-hosting.dashnexpages.net/thylandpublishing/open-access-articles/tijocs-jse-20200207-516.pdf
45. Johri, A., Kumar, S.: Exploring customer awareness towards their cyber security in the Kingdom of Saudi Arabia: a study in the era of banking digital transformation. Hum. Behav. Emerg. Technol. **2023** (2023). https://doi.org/10.1155/2023/2103442
46. Shakhov, V., Koo, I.: Graph-based technique for survivability assessment and optimization of IoT applications. Int. J. Softw. Tools Technol. Transf. **23**(1), 105–114 (2021). https://doi.org/10.1007/s10009-020-00594-9
47. Mishra, N., Pandya, S.: Internet of things applicaations, security challenges, attacks, intrusion detection, and future visions: a systematic review. IEEE Access **9**, 59353–59377 (2021). https://doi.org/10.1109/ACCESS.2021.3073408
48. Butun, I., Osterberg, P., Song, H.: Security of the internet of things: vulnerabilities, attacks, and countermeasures. IEEE Commun. Surv. Tutorials **22**(1), 616–644 (2020). https://doi.org/10.1109/COMST.2019.2953364
49. Alasan, A.I., Eyanuku, J.P.: Human resource management practices and nigerian banks' performance. LASU. J. Emp. Relat. Hum. Resour. Manag. **2**(1), 124–140 (2020). https://doi.org/10.36108/ljerhrm/0202.02.0190
50. Madanaguli, A., Srivastava, S., Ferraris, A., Dhir, A.: Corporate social responsibility and sustainability in the tourism sector: a systematic literature review and future outlook. Sustain. Dev. 1–15 (2021). https://doi.org/10.1002/sd.2258
51. Fahim, I., Poursalimi, M., Hosseinzadeh, A., Namaghi, M.G.: Evaluation of a branding model with emphasis on organizational social responsibility based on social networks in the banking industry and the structural equation method. Cogent Bus. Manag. **8**(1) (2021). https://doi.org/10.1080/23311975.2021.1908744
52. Khanagha, S., Volberda, H.W., Alexiou, A., Annosi, M.C.: Mitigating the dark side of agile teams: Peer pressure, leaders' control, and the innovative output of agile teams. J. Prod. Innov. Manag. **39**(3), 334–350 (2022). https://doi.org/10.1111/jpim.12589
53. Malinka, K., Hujnak, O., Hanacek, P., Hellebrandt, L.: E-banking security study-10 years later. IEEE Access **10**, 16681–16699 (2022). https://doi.org/10.1109/ACCESS.2022.3149475

Chapter 5
Best Practices for Mitigating Risks, Conclusion and Recommendation

5.1 Best Practices for Mitigating Risks

In the investigation titled "The Implication of Cyberattacks on Big Data and How to Mitigate the Risk," it is discovered that the combination of IAM, Network Segmentation, and SDE algorithms constitutes a powerful line of defense against the ever-increasing risk posed by cyberattacks. This chapter has shed light on the crucial need to preserve valuable assets related to Big Data and offered a thorough strategy for mitigating risks associated with these assets. The previous chapter emphasized the necessity of implementing stringent cybersecurity precautions by digging into the various ramifications that may result from a cyberattack on Big Data. To successfully handle these difficulties, a multi-dimensional approach that combines IAM, Network Segmentation, and SDE algorithms has emerged as an option [1].

IAM's function regulates user access and permissions and is the primary defensive mechanism. This integration, in conjunction with Network Segmentation, which separates the network into autonomous portions, prevents illegal lateral movement and controls possible breaches. SDE provides an additional degree of security by making the data unintelligible to any third parties who are not allowed to access it [2]. This all-encompassing strategy significantly minimizes the organization's vulnerability to cyber-attacks, improves data confidentiality, and strengthens the business's resistance to external and internal acts of harmful activity [3]. While it is understood that there will be hurdles, such as complicated implementation, the advantages that will arise are enormous. When combined, IAM, Network Segmentation, and SDE algorithms provide a strengthened cybersecurity approach, which is crucial in the digital era [4]. In this thesis, we have traversed the landscape of the cybersecurity dangers connected with Big Data, and we have shown a certain approach to protecting the integrity, privacy, and accessibility of priceless data assets.

This book design presents a thorough approach to exploring the implications of cyberattacks on big data and generating strategies to mitigate the related risks. With

F. A. Alaba and A. Rocha, *The Implication of Cyberattacks on Big Data and How to Mitigate the Risk*, Information Systems Engineering and Management 39,
https://doi.org/10.1007/978-3-031-88570-9_5

a particular emphasis on IAM, Symmetric Data Encryption, and Network Segmentation, the plan will investigate the implications of cyberattacks on big data and create methods to mitigate associated risks [5]. The book intends to accomplish several goals, including examining the function of Network Segmentation, determining the impact of cyberattacks on big data, determining the efficacy of IAM and encryption in safeguarding big data and presenting complete techniques for mitigating the effects of cyberattacks [6].

The book will use a mixed-methods approach, which will include collecting and examining qualitative and quantitative data, respectively. During the qualitative phase of the book, in-depth interviews with cybersecurity professionals and a review of relevant literature will be conducted to give insights into the book topic. In the quantitative phase, we will collect quantitative data for analysis by polling IT and data security specialists through digital survey forms. The participants will be selected via sampling procedures, to assure the presence of experts and professionals with relevant experience in big data security. Ethical concerns will be of the utmost importance, and safeguards will be in place to preserve the privacy and confidentiality of participants.

The book of the data will include both theme analysis for qualitative data and statistical approaches for quantitative data, so reliable conclusions can be drawn. The reliability and scientific validity of the book will be improved if the appropriate steps are taken. This book design considers possible restrictions, such as the availability of participants and the reliability of data that participants report on their own, and it provides a flexible timeframe and budget to account for unforeseen circumstances. In the end, the purpose of this book is to make a significant contribution to the field of cybersecurity by investigating the multifaceted problem of big data security, putting forward efficient mitigation strategies, and advancing our understanding of the roles that identity and access management, symmetric data encryption, and network segmentation play in the process of protecting big data environments [7].

To address the potential hazards stemming from cyberattacks on Big Data, organisations need to embrace a comprehensive and proactive strategy. The implementation of several best practices is necessary in this context. These practises encompass data encryption, access control, regular auditing and monitoring, patch management, intrusion detection systems, incident response plans, employee training, third-party security, backups, network segmentation, data masking, legal and compliance considerations, collaboration, regular security assessments, and continuous improvement. The use of data encryption is of utmost importance in safeguarding data both during storage and while being sent. By using strong encryption methods, organisations may effectively safeguard their data from unauthorized access by cybercriminals. This is achieved by rendering the data unintelligible without the corresponding encryption keys, so ensuring its confidentiality and integrity. Access control is crucial practise in Big Data repositories, as it restricts access only to authorised individuals. This entails the implementation of stringent access controls and user authentication procedures [8].

Implementing routine auditing and monitoring procedures is crucial to identify any irregularities and risks effectively. It is recommended that organisations establish alert mechanisms for detecting and monitoring anomalous actions while also conducting regular security audits to protect the integrity of their systems. The administration of software patches is of utmost importance, as it is crucial in ensuring that all software is regularly updated with the newest security patches. This proactive approach significantly reduces the risk of potential vulnerabilities being exploited. Using intrusion detection systems enables organisations to identify and address possible security risks or atypical network activities. Implementing a comprehensive incident response strategy, accompanied by frequent testing, enhances an organization's preparedness to efficiently managing cyberattacks. The significance of employee training lies in its ability to provide knowledge to workers on optimal practises for online conduct, enabling them to identify and respond to phishing efforts and other forms of social engineering [9].

When organisations engage in collaborations with third-party vendors or cloud service providers, it is imperative that they ascertain the adherence of these entities to rigorous security requirements and their ability to effectively safeguard sensitive data. The implementation of routine backup procedures for Big Data repositories, along with the storage of these backups in a safe place, facilitates the process of data restoration in the case of a cyberattack. The act of segmenting the network serves to provide isolation between Big Data environments and other components of the network, hence enhancing the level of security. Using data masking strategies safeguards sensitive data, and organizations must ensure that their cybersecurity practices follow legal and regulatory obligations [10]. Collaborating with the cybersecurity community facilitates the acquisition of current knowledge on new risks and optimal methodologies, enabling organizations to remain well-informed and up-to-date. Regular security assessments and penetration testing identify vulnerabilities and weaknesses within the Big Data architecture, enabling prompt correction. Continuous improvement is vital in cybersecurity due to its dynamic nature, where the need to respond and adapt to ever-changing threats and technology is constant [11].

By adhering to these recommended methodologies, enterprises may effectively mitigate the vulnerabilities linked to assaults on Big Data, safeguarding their data's integrity, confidentiality, and availability.

5.1.1 Implementing a Comprehensive Cybersecurity Policy

A comprehensive cybersecurity strategy is paramount to protecting an organization's digital assets and data. Presented below is a comprehensive, sequential framework aimed at facilitating the development and implementation of a robust cybersecurity policy [11–14].

1. Stakeholder Identification: Identify primary stakeholders and decision-makers accountable for formulating, executing, and upholding the policy.
2. Risk Assessment: Perform a comprehensive risk assessment to ascertain possible vulnerabilities and threats, including both internal and external factors.

3. strategy Development: Formulate an all-encompassing cybersecurity strategy that effectively tackles the distinct requirements and vulnerabilities of the organisation. The essential topics that should be addressed are access control, safeguarding of data, responding to incidents, adherence to acceptable use policies, administration of passwords, ensuring network security, overseeing mobile device use, facilitating remote work, providing staff training, ensuring vendor security, compliance with regulations, management of Bring Your Own Device (BYOD) policies, securing cloud environments, and mitigating social engineering threats.
4. Policy Review and Approval: Engage relevant stakeholders in the process of examining and endorsing the policy to guarantee its congruence with organisational objectives and regulatory obligations.
5. Communication and Training: Disseminate the policy to all personnel and provide comprehensive training sessions to ensure comprehension and adherence to its stipulations. It is important to ensure that personnel possess a comprehensive understanding of their respective roles and the potential ramifications associated with any infractions of organisational policies.
6. Implementation of Enforcement and Monitoring Measures: Deploy several techniques to effectively monitor and enforce the policy, including the use of access control tools, intrusion detection systems, and conducting periodic security assessments.
7. Development of an Elaborate Incident Response Plan: Construct a comprehensive incident response plan delineating the sequential actions undertaken during a security issue. It is important to ensure that personnel possess a comprehensive understanding of the protocols for reporting occurrences and the processes pertaining to containment, investigation, and recovery.
8. Consistent Updates: It is important to conduct regular assessments and policy revisions to effectively respond to the ever-changing landscape of cyber threats and new technologies.
9. Testing and Evaluation: It is important to regularly evaluate security measures to ascertain their efficacy and employee adherence to these controls.
10. Compliance: It is important to verify that the policy is following relevant industry standards and regulatory obligations, including but not limited to GDPR, HIPAA, and other applicable legislation.
11. Ongoing Enhancement: Facilitate the continual improvement of the policy by acquiring knowledge from security incidents and adhering to industry-leading standards.
12. Legal and Ethical Considerations: This section examines the legal and ethical dimensions of the study, including topics such as privacy, consent, and adherence to international data protection rules.

Organizations may establish a robust framework for safeguarding their digital assets and data from cyberattacks by adhering to these prescribed procedures and consistently evaluating and adjusting their cybersecurity strategy.

5.1.2 Regular Security Audits and Vulnerability Assessments

Establishing and implementing a comprehensive cybersecurity strategy is crucial in safeguarding an organization's digital assets and confidential data. To do this task proficiently, it is advisable to adhere to the following sequential instructions [15, 16].

1. Formulate a cross-functional team: Convene a collective of information technology specialists, legal professionals, compliance officials, and key decision-makers to devise and execute the policy above.
2. Perform a comprehensive risk assessment: Identify and evaluate possible cybersecurity vulnerabilities and threats unique to your organisation, including internal and external factors.
3. Formulate the policy: Establish an all-encompassing cybersecurity policy that encompasses various aspects, including access control, data safeguarding, incident response, acceptable usage, password administration, network protection, management of mobile devices, remote work protocols, employee training, vendor security, compliance measures, bring-your-own-device (BYOD) policies, cloud security, and countermeasures against social engineering.
4. Policy Evaluation and Authorization: Submit the policy to relevant stakeholders for evaluation and authorization, confirming its congruence with organisational objectives, industry norms, and regulatory obligations.
5. Facilitate communication and training: Employ efficient communication strategies to disseminate the policy to all personnel and conduct comprehensive training sessions to guarantee comprehension of its stipulations and the potential repercussions for non-compliance.
6. Implementation and Oversight: Establish and execute mechanisms and protocols for the enforcement and monitoring of the policy, including the use of access control mechanisms, intrusion detection systems, and periodic security assessments.
7. Formulate an incident response plan: Elaborate a comprehensive strategy delineating the sequential actions to be undertaken in the event of a security incident, including the aspects of reporting, containment, investigation, and recovery.
8. It is important to regularly update the policy by consistently reviewing and revising it in order to effectively meet evolving threats and technology.
9. Testing and evaluation: Regularly perform assessments and evaluations of security measures to ascertain their efficacy and adherence by employees.
10. Ensure regulatory adherence: Establish congruence between the policy and relevant industry benchmarks and legal obligations, such as the General Data Protection Regulation (GDPR) or the Health Insurance Portability and Accountability Act (HIPAA).
11. The policy should be subject to continual improvement by learning from security events and keeping abreast of industry best practises.
12. The policy should take into account the legal and ethical dimensions, specifically focusing on privacy, consent, and adherence to international data protection rules.

By adhering to the prescribed procedures, entities may create and implement a comprehensive cybersecurity protocol aimed at safeguarding their digital resources and confidential data from potential cyber hazards.

5.1.3 Collaborative Efforts with Cybersecurity Experts

The involvement of cybersecurity professionals in collaborative endeavours is crucial for bolstering the security stance of your organisation and efficiently addressing cyber hazards. The following are a series of measures that may be implemented to cultivate cooperation with professionals in the field of cybersecurity.

1. Recognise the Importance of Teamwork: Identify the precise domains within your organisational structure necessitating engagement with cybersecurity professionals. Possible areas of focus within the realm of cybersecurity including threat analysis, incident response, security audits, and policy formulation.
2. The Cybersecurity Group Within: It is recommended to establish an internal cybersecurity team in the event that one is not already in place. It is recommended that the composition of this team consist of individuals with professional qualifications and specialised knowledge in the fields of cybersecurity, information security, network security, and compliance.
3. Find Outside Cybersecurity Professionals: The task at hand involves the identification and establishment of links with external specialists and organisations in the field of cybersecurity. Potential professionals in the field of cybersecurity may include security consultants, penetration testers, incident response businesses, and legal advisers with a specialisation in cybersecurity.
4. Express Yourself Clearly: Effectively convey your organization's cybersecurity requirements and objectives to the external specialists. Furnish them with a full comprehension of your company procedures, systems, and data in order to facilitate efficient cooperation.
5. Recovery from Disruptions: Engage in a collaborative effort with external subject matter experts to develop an incident response plan that include communication protocols, delineation of roles, and assignment of tasks in the event of a security issue. This inquiry seeks to ascertain the manner in which external experts will provide assistance in the case of a breach.
6. Conducting Frequent Reviews of Security: It is advisable to enlist the services of external cybersecurity professionals in order to carry out routine security assessments, which include vulnerability assessments, penetration testing, and security audits.
7. Formulation of Policy: Engage in collaborative efforts with industry professionals to establish and enhance cybersecurity policies, standards, and optimal methodologies inside your organisation.

8. Education and Knowledge: Collaborate with industry professionals in order to develop and implement cybersecurity training and awareness initiatives aimed at enhancing workers' adherence to secure practises.
9. Collaboration on Threat Intelligence: Facilitate the sharing of threat information with external subject matter experts and peers inside the sector. The act of exchanging knowledge may facilitate the ability to predict and safeguard against new hazards.
10. Compliance with Requirements: It is advisable to engage in collaboration with cybersecurity professionals in order to guarantee that your organisation adheres to pertinent industry rules and data protection legislation.
11. Evaluating New Technologies: It is advisable to consult with professionals in the field while assessing and choosing cybersecurity technology and solutions. They possess the ability to assist in the identification of the most appropriate tools for the unique requirements of your organisation.
12. Discuss safe methods for creating software: To guarantee the integration of security throughout the software development lifecycle, it is advisable for organisations engaged in software development to collaborate with professionals in the field.
13. Modelling and Testing of Incidents: To enhance an organization's preparedness for a cybersecurity breach, it is recommended to engage in cybersecurity incident simulations and tabletop exercises in collaboration with experts. These activities aim to simulate potential breach scenarios and evaluate the organization's response capabilities. By doing such exercises, organisations may assess their preparation, identify any gaps or weaknesses in their incident response plans, and develop strategies to mitigate and address such breaches effectively.
14. The assessment of vendor and third-party risks: To lessen the impact of potential supply chain disruptions, businesses should work with cybersecurity professionals to evaluate the vendor and supplier community's security procedures.
15. Ethics and the Law: It is advisable to consult with cybersecurity professionals with expertise in legal and ethical matters to verify that your cybersecurity practices adhere to legal and ethical standards. International data protection rules have significant importance in this context.
16. Enhancement In perpetuity: It is important to maintain ongoing collaboration with cybersecurity specialists to remain current with threats and adhere to established best practices. It is essential to constantly revise and enhance security protocols in order to mitigate newly emerging hazards effectively.
17. Create an Environment of Openness and Trust: Cultivate a rapport characterized by trust and transparent communication with external specialists. Collaboratively enhance the security stance of your organisation by exchanging knowledge and valuable perspectives.

The continuous engagement with cybersecurity professionals facilitates organisational adaptation to the dynamic threat environment. These esteemed professionals provide unique perspectives, expertise, and practical know-how to bolster your organization's cybersecurity resilience and safeguard against cyber attacks.

5.1.4 Incident Response Plans and Crisis Management

When it comes to proactively addressing cybersecurity risks and ensuring the continuation of business, it is very necessary for organisations to develop and put into action comprehensive incident response plans and crisis management. The most important stages are as follows:

1. Put together an emergency response team: Develop a specialised staff with experience in areas such as cyber security, legal issues, public relations, and business operations.
2. Specify the Types of Incidents: Determine the possible security events that might have an effect on your company, such as data breaches, malware infections, distributed denial of service attacks, or threats from inside the organisation.
3. Create an event Response strategy: Develop a detailed strategy that specifies the measures to follow when a security event happens, covering multiple incident kinds and situations. This step is the third stage in the process.
4. Incident Identification and Reporting: Establish protocols for rapidly identifying and reporting security events, making sure that all workers are aware of how to report occurrences, and set a deadline for when these procedures must be completed.
5. Classification and Prioritisation: Create a system for the classification and prioritisation of events depending on the severity of the occurrences and the possible effect they might have on the organisation.
6. Outline thorough response processes for each kind of event, including as containment, threat elimination, system recovery, and communication with stakeholders.
7. Formulate a strategy for both internal and external communication methods, ensuring that all workers, management, customers, and regulatory agencies are kept correctly informed. This is the seventh and last step in the process.
8. Containment and Eradication: Define the procedures that will be taken in order to limit the event and eliminate the danger, such as isolating the systems that have been compromised, uninstalling any malware, and recovering any lost data.
9. Recovery and Restoration: Describe the procedures that need to be taken in order to recover from the event, including the restoration of the system, the recovery of the data, and a return to regular operations.

10. Compliance with Relevant Laws and Regulations Make certain that the incident response plan is in accordance with all applicable legal and regulatory obligations, such as those pertaining to the disclosure of data breaches.
11. Testing and Training: The incident response plan should be tested on a regular basis using simulations and exercises. In addition, the response team should get training to ensure that everyone is comfortable with their respective roles and responsibilities.
12. Maintain and keep up to date the incident response strategy in order to account for changes in the threat environment, infrastructure, or regulatory requirements.
13. Crisis Management: Incorporate incident response into a more comprehensive framework for crisis management. Establish communication channels and processes for handling reputational harm and public relations.
14. Continuity of Business: Check to see that your incident response procedures are in line with your larger continuity of business and catastrophe recovery policies.
15. Public Relations: Maintain consumer trust and safeguard the organization's image by working closely with the public relations team to handle external communications.
16. Documentation and Analysis: Keep complete records of events and responses for post-incident analysis; use the data to refine the response plan and suggest areas for improvement.
17. Conduct in-depth assessments and analyses of each event after it has been resolved in order to determine how successful the reaction was and what can be learnt from it.

5.2 Conclusion

In conclusion, in this book, I have conducted an in-depth analysis of cyberattacks' crucial implications on the availability, integrity, and confidentiality of Big Data assets. The necessity for comprehensive cybersecurity measures is more apparent now than ever before due to the rising complexity of digital environments. Throughout this chapter, we dove deep into the many issues presented by cyber threats and their ability to jeopardise the priceless Big Data resources that businesses have at their disposal. A comprehensive and efficient strategy for reducing these threats is the combination of IAM, Network Segmentation, and SDE algorithms. When combined with Network Segmentation's function in isolating key assets, IAM's role in controlling user access and permissions creates a powerful barrier against illegal access and lateral movement inside the network. The IAM establishes this defense. SDE assures that even if data is read, it does not become understandable to unauthorised parties. This is an additional degree of protection that SDE provides. The combined strategy of IAM, Network Segmentation, and SDE utilises their respective capabilities to create a powerful cybersecurity architecture that protects against various attacks. While it is true that the execution of this connection presents certain problems, such as those related to complexity and the user experience, it is also abundantly clear that the advantages of this integration considerably exceed the challenges.

In a world in which the possibility of cyberattacks is always increasing, the implementation of integrated identity and access management, network segmentation, and sensitive data encryption techniques allow businesses to safeguard their most important big data asset holdings. This chapter gives a road map for enterprises to follow to successfully cross the difficult terrain of cybersecurity threats while maintaining the availability, integrity, and confidentiality of their essential information resources.

Furthermore, this book design presents a complete method for examining the consequences of cyberattacks on big data and devising effective measures to manage the risks that are connected with these assaults. The book emphasises IAM, Symmetric Data Encryption, and Network Segmentation. This reflects the changing landscape of cybersecurity and the important requirement to safeguard sensitive data in settings containing large amounts of data. This book design intends to give a comprehensive knowledge of the issues presented by cyber threats to big data by creating explicit book goals and applying a mixed-methods approach that integrates qualitative and quantitative data. In other words, it does this by setting defined book objectives. The qualitative phase, which will include interviews and literature studies, will provide very helpful insights into the experiences and points of view of professionals in the field of cybersecurity. In the meanwhile, the quantitative phase, which will consist of surveys administered to IT experts and data security specialists, will collect quantifiable data for the purpose of conducting in-depth analysis. The design of the book places a strong emphasis on ethical issues, with the goal of preserving the participants' right to privacy and adhering to established ethical standards. To ensure that the results of this book continue to be of a high quality and preserve their credibility, we have devised criteria for data analysis, validation, and dependability. This book design provides a solid framework for efficiently addressing the book goals, in spite of the fact that it acknowledges the possibility of some constraints and the need for flexibility in both the timeframe and the budget.

In summary, the purpose of this book design is to make substantial contributions to the area of cybersecurity by providing insights into the consequences of cyberattacks on big data and proposing complete mitigation solutions that utilize identity and access management, symmetric data encryption, and network segmentation. In a world that is becoming more digital and linked, the results of this book have the potential to contribute to the improvement of cybersecurity procedures and to the preservation of the integrity and safety of big data assets.

5.3 Recommendation

Several recommendations for improving cybersecurity arise from the in-depth investigation of "The Implication of Cyberattacks on Big Data and How to Mitigate the Risk" via integrated IAM, Network Segmentation, and SDE algorithms [17–19].

- It is strongly recommended that businesses use IAM, Network Segmentation, and SDE to achieve a comprehensive strategy for cybersecurity. Your computer will have enough protection against cyber assault if you use these three layers of security [20].
- The effectiveness of the implementation is dependent on regular training and awareness campaigns that remind staff members of the relevance of cybersecurity and their role in protecting it. Employees may be equipped to recognize security hazards and respond appropriately if there is consistent training and awareness training [21].
- Because cyber threats are always evolving, keeping an eye on them is essential. Install a round-the-clock surveillance system to identify and plug any new security flaws. It is necessary to conduct frequent evaluations to determine how well the integrated approach operates and make any necessary adjustments [22].
- Because maintaining cybersecurity requires the combined efforts of many people, each individual is responsible for doing their share. Encourage communication between IT departments, high management, and staff members to guarantee that security measures are properly installed and routinely maintained [23].
- Create a comprehensive plan (also known as an "incident response plan") that explains what steps need to be taken in the event of a breach in cybersecurity. This plan must include strategies for locating the problem, containing it, doing away with it, and recovering from its effects [24].
- When working with different suppliers, confirming that their safety practices are comparable to your own is essential. Demand that there be transparency on safety practices as well as frequent updates [25].
- Create backups of your data consistently and store them in a secure location. If your data is compromised, having current backups may help you retrieve your files more quickly and minimize the lost data [26].
- Because threats to cybersecurity and recommendations for best practices are always evolving, it is essential to stay current. Make advantage of the resources provided by the industry and the events held by the industry to get knowledge about emerging dangers, viable countermeasures, and best practices [27].
- Make it a point to check that the integrated approach conforms with all relevant industry standards and legislation around data protection and privacy. Keeping up with compliance requirements boosts security and helps critical participants feel more confident in the system [28].
- The importance of cyber security should be emphasized across a whole firm, but most significantly at the executive level. When upper management takes the issue of security seriously, workers tend to do the same [29].
- Consider enlisting the assistance of specialists in the field of cybersecurity to do assessments, audits, and recommendations. Their expertise could be useful in identifying security flaws and making the system more foolproof [30].

Following the guidance presented above is highly recommended for companies that prioritise maintaining the safety of their Big Data holdings in the face of the rapid evolution of the digital world. Businesses significantly reduce their susceptibility to cyberattacks and preserve the confidentiality of their sensitive information when they take these preventative measures and put them into practice.

References

1. Kehayov, M., Holder, L., Koch, V.: Application of artificial intelligence technology in the manufacturing process and purchasing and supply management. Proc. Comput. Sci. **200**(2019), 1209–1217 (2022). https://doi.org/10.1016/j.procs.2022.01.321
2. Al-Garadi, M.A., Mohamed, A., Al-Ali, A.K., Du, X., Ali, I., Guizani, M.: A survey of machine and deep learning methods for internet of things (IoT) security. IEEE Commun. Surv. Tutorials **22**(3), 1646–1685 (2020). https://doi.org/10.1109/COMST.2020.2988293
3. Çetin, S., De Wolf, C., Bocken, N.: Circular digital built environment: an emerging framework. Sustainability **13**(11), 1–34 (2021). https://doi.org/10.3390/su13116348
4. Kumar, A.A.D., Kusonthammarat, P., Guzman, A.L., Zohuri, B.: Supply chain driven supply and demand augmenting resiliency integrated artificial intelligence. J. Econ. Manag. Res. **3**(1), 1–4 (2022). https://doi.org/10.47363/jesmr/2022(3)146
5. Maria Tsikala Vafea, E.M., Atalla, E., Georgakas, J., Shehadeh, F., Mylona, E.K., Kalligeros, M.: Emerging technologies for use in the study , diagnosis, and treatment of patients with COVID-19. Biomed. Eng. Soc. **13**(4), 249–257 (2020). https://doi.org/10.1007/s12195-020-00629-w
6. Zimba, A.: A Bayesian attack-network modeling approach to mitigating malware-based banking cyberattacks. Int. J. Comput. Netw. Inf. Secur. **14**(1), 25–39 (2022). https://doi.org/10.5815/ijcnis.2022.01.03
7. Seungjin, L., Abdullah, A., Jhanjhi, N.Z.: A review on honeypot-based botnet detection models for smart factory. Int. J. Adv. Comput. Sci. Appl. **11**(6), 418–435 (2020). https://doi.org/10.14569/IJACSA.2020.0110654
8. Al Qartah, A.: Evolving ransomware attacks on healthcare providers. ICASSP, IEEE International Conference on Acoustics, Speech, and Signal Processing—Proceedings, vol. 01, no. August, pp. 12–24 (2020). https://doi.org/10.13140/RG.2.2.23202.45765
9. Safarov, F., Basak, M., Nasimov, R., Abdusalomov, A., Cho, Y.I.: Explainable lightweight block attention module framework for network-based IoT attack detection. Future Internet **15**(9), 297 (2023). https://doi.org/10.3390/fi15090297
10. Torres, N., Pinto, P., Lopes, S.I.: Security vulnerabilities in LPWANs—an attack vector analysis for the IoT ecosystem. Appl. Sci. **11**(7), 3176 (2021). https://doi.org/10.3390/app11073176
11. Dong, T., Li, S., Qiu, H., Lu, J.: An Interpretable Federated Learning-based Network Intrusion Detection Framework, pp. 1–12 (2022). http://arxiv.org/abs/2201.03134
12. Hampton, N., Baig, Z., Zeadally, S.: Ransomware behavioural analysis on windows platforms. J. Inf. Secur. Appl. **40**, 44–51 (2018). https://doi.org/10.1016/j.jisa.2018.02.008
13. Panoff, M., Dutta, R.G., Hu, Y., Yang, K., Jin, Y.: On Sensor security in the era of IoT and CPS. SN Comput. Sci. **2**(1), 1–14 (2021). https://doi.org/10.1007/s42979-020-00423-5
14. Krichen, M.: Strengthening the security of smart contracts through the power of artificial intelligence. Computers **12**(5), 1–18 (2023). https://doi.org/10.3390/computers12050107
15. Jiang, Y., Atif, Y.: A selective ensemble model for cognitive cybersecurity analysis. J. Netw. Comput. Appl. **193**(September), 103210 (2021). https://doi.org/10.1016/j.jnca.2021.103210
16. Gupta Gourisetti, N., Mylrea, M., Patangia, H.: Application of rank-weight methods to blockchain cybersecurity vulnerability assessment framework. In: 2009 IEEE 9th Annual Computing and Communication Workshop and Conference, CCWC 2019, pp. 206–213 (2019). https://doi.org/10.1109/CCWC.2019.8666518
17. Malinka, K., Hujnak, O., Hanacek, P., Hellebrandt, L.: E-banking security study-10 years later. IEEE Access **10**, 16681–16699 (2022). https://doi.org/10.1109/ACCESS.2022.3149475

18. Orabi, M., Mouheb, D., Al Aghbari, Z., Kamel, I.: Detection of bots in social media: a systematic review. Inf. Process. Manag. **57**(4) (2020). https://doi.org/10.1016/j.ipm.2020.102250
19. Golightly, L., Chang, V., Xu, Q.A., Gao, X., Liu, B.S.C.: Adoption of cloud computing as innovation in the organization. Int. J. Eng. Bus. Manag. **14**, 1–17 (2022). https://doi.org/10.1177/18479790221093992
20. Pambhar, H., Aghera, K., Tada, N., Residual, Á.Á.: An advanced web-based bilingual domain independent interface to database using machine learning approach **508**(June), 197–204 (2018). https://doi.org/10.1007/978-981-10-2750-5
21. Ohri, K., Kumar, M.: Review on self-supervised image recognition using deep neural networks. Knowl. Based Syst. **224**, 107090 (2021). https://doi.org/10.1016/j.knosys.2021.107090
22. Baskaran, K.R.: Deep learning based early Diagnosis of Alzheimer's disease using semi supervised GAN. Ann. Rom. Soc. Cell Biol. **25**(3), 1583–6258 (2021). http://annalsofrscb.ro
23. Xu, G., Jin, H.: Using artificial intelligence technology to solve the electronic health service by processing the online case information. J. Healthc. Eng. **2021**(3), 1–12 (2021). https://doi.org/10.1155/2021/9637018
24. Comelli, A., et al.: Deep learning-based methods for prostate segmentation in magnetic resonance imaging. Appl. Sci. **11**(2), 1–13 (2021). https://doi.org/10.3390/app11020782
25. Hassan, B., et al.: Deep learning based joint segmentation and characterization of multi-class retinal fluid lesions on OCT scans for clinical use in anti-VEGF therapy. Comput. Biol. Med. **136**(July), 104727 (2021). https://doi.org/10.1016/j.compbiomed.2021.104727
26. Healthcare, A.I., Awan, K.A., Din, I.U., Almogren, A., Khattak, H.A., Rodrigues, J.J.P.C.: EdgeTrust : A Lightweight Data-Centric Trust Management, pp. 1–20 (2023)
27. Mohammad Shah, I.N., Ismail, E.S., Samat, F., Nek Abd Rahman, N.: Modified generalized feistel network block cipher for the internet of things. Symmetry **15**(4) (2023). https://doi.org/10.3390/sym15040900
28. Saad, M., Bin Ahmad, M., Asif, M., Khan, M.K., Mahmood, T., Mahmood, M.T.: Blockchain-enabled VANET for smart solid waste management. IEEE Access **11**(November), 5679–5700 (2023). https://doi.org/10.1109/ACCESS.2023.3235017
29. Li, X., Liu, S., Kumari, S., Chen, C.M.: PSAP-WSN: a provably secure authentication protocol for 5G-based wireless sensor networks. Comput. Model. Eng. Sci. **135**(1), 711–732 (2023). https://doi.org/10.32604/cmes.2022.022667
30. Tekinerdogan, B., Köksal, Ö., Çelik, T.: System architecture design of IoT-based smart cities. Appl. Sci. **13**(7) (2023). https://doi.org/10.3390/app13074173

The manufacturer's authorised representative in the EU is Springer Nature Customer Service Centre GmbH, Europaplatz 3, 69115 Heidelberg, Germany. If you have any concerns regarding our products, please contact ProductSafety@springernature.com

Printed and bound by CPI Group (UK) Ltd, Croydon, CR0 4YY

01/10/2026

02225977-0006